CONFERENCE PROCEEDINGS

Edited by

Dr. Sai Preethi Prasath
Head, Department of Biotechnology
Sree Sastha Institute of Engineering and Technology,
Chembarambakkam, Chennai - 600123.

PUBLISHERS DETAILS

On behalf of

Sree Sastha Institute of Engineering and Technology

Published by

Dr. Shanmuga Sundaram Anandan,
Principal & Director,
Sree Sastha Institute of Engineering and Technology,
Chembarambakkam, Chennai – 600123.

INTERNATIONAL CONFERENCE ON
BIOPHARMACEUTICALS, NANOBIOTECHNOLOGY & NANOSCIENCES (ICBNN- 2025)

ISBN: 978-93-341-8379-5

Conference Proceedings,

January 10, 2025 by

Dr. Shanmuga Sundaram Anandan,
Principal & Director,
Sree Sastha Institute of Engineering and Technology,
Chembarambakkam, Chennai - 600123.

Printed and bound in India by
Qurate Books Pvt Ltd, Panaji, Goa-403001.
info@quratebooks.com

FOREWORD BY CHAIRMAN

It gives me great pleasure to welcome all participants of the International Conference on Biopharmaceuticals, Nanobiotechnology & Nanosciences (ICBNN-2025) organized by Department of Biotechnology, Sree Sastha Institute of Engineering and Technology (SSIET), Chembarambakkam. I would also like to congratulate the committee members from the Department of Biotechnology for taking up the challenge of hosting this event.

As a quality, engineering and technology institution, SSIET needs to exploit and leverage upon all possible areas of opportunities and outputs. Through conferences and seminar such as ICBNN-2025, research outputs can always be presented and later realized as publications that serve to enrich the existing body of knowledge of this field. Interactions during the conference can further stimulate research ideas, provide opportunities for research co-operation and enable the stakeholders to network and exchange information and knowledge. With ICBNN-2025 focusing on Biopharmaceuticals, Nanobiotechnology & Nanosciences, it also plays the role of transforming abstract concepts into useful technology and, as such, preserves and expands the innovation. It is hoped that individual researchers, particularly those from SSIET will work hand in hand, in developing this important field of research and pushing it together to be the most important vehicle for value creation within the whole society.

I would also like to congratulate all the researchers who have achieved their desired results in their research work, and I hope their findings will be translated into practical use in the relevant commercial fields, thus benefiting our nation and others beyond our borders.

Finally, I would like to express my sincere thanks to the organizing committee from Department of Biotechnology for their efforts in organizing this conference. To all participants, wishing you a successful conference.

Thank you.

With Regards
Prof. J. Kartheekeyan
Chairman
Sree Sastha Group of Institutions.
Chembarrambakkam, Chennai-600123.

FOREWORD BY PRINCIPAL

I am deeply honored and privileged to warmly welcome all participants, esteemed speakers, and delegates attending the International Conference on Biopharmaceuticals, Nanobiotechnology, and Nanosciences (ICBNN-2025). Hosted by the Department of Biotechnology at Sree Sastha Institute of Engineering and Technology, this conference provides a leading platform for intellectual exchange, innovation advancement, and collaboration in some of today's most dynamic fields of scientific research.

The intersection of biopharmaceuticals, nanotechnology, and nanosciences is propelling ground breaking advancements, revolutionizing industries and tackling some of humanity's most urgent issues. This conference aims to enhance these efforts by gathering thought leaders, researchers, and practitioners to exchange their expertise, insights, and innovative discoveries.

At Sree Sastha Institute of Engineering and Technology, we are dedicated to promoting excellence in research and education. Events like ICBNN-2025 highlight our vision of cultivating a dynamic academic environment that enables individuals and institutions to innovate, collaborate, and create solutions for the betterment of society worldwide.

I applaud the organizing committee for their careful work in planning this event and bringing together an impressive group of speakers and attendees. I am sure that the discussions and exchanges during this conference will spark new ideas, drive innovative solutions, and create lasting collaborations across various fields.

I urge all participants to actively contribute, engage in thoughtful conversations, and fully embrace this exceptional opportunity. Let this conference be a source of inspiration and advancement for everyone involved.

Wishing the best of success to ICBNN 2025! Thank you.

With Regards
Dr.Shanmuga Sundaram Anandan
Principal & Director
Sree Sastha Group of Institutions.
Chembarrambakkam, Chennai-600123.

FOREWORD BY EDITOR

Welcome to International Conference on Biopharmaceuticals, Nanobiotechnology & Nanosciences (ICBNN-2025). It is my great pleasure to welcome all of you to the International Conference on Biopharmaceuticals, Nanobiotechnology & Nanosciences (ICBNN-2025) organized by Department of Biotechnology, Sree Sastha Institute of Engineering and Technology (SSIET), Chembarambakkam.

International Conference on Biopharmaceuticals, Nanobiotechnology & Nanosciences (ICBNN-2025) is aimed at providing the scientists, researchers and engineers from all over India a scientific forum of discussions to exchange ideas and the research results in the fields of Biopharmaceuticals, Nanobiotechnology & Nanosciences, as to develop new scientific discoveries in these area of interests. We have around 200 participants in the conference. The number of participants in the oral presentations was 49, while 27 participants took part in the poster presentations.

With respect to the hard work of the committee, I would like to express my deep appreciation and gratitude to all the committee members, the ICBNN-2025 committee members whom have made this conference a success. And especially to the Department of Biotechnology, Sree Sastha Institute of Engineering and Technology (SSIET), Chembarambakkam that has supported the whole effort in making the conference a success.

Last but not least, I would like to thank all the keynote speakers and the plenary speakers, and fellow researchers that has been willing to share their research knowledge and experience to all of us in this prestigious conference.

Thank you.

With Regards
Dr.Sai Preethi Prasath
Head, Department of Biotechnology
Sree Sastha Institute of Engineering and Technology,
Chembarambakkam, Chennai - 600123.

ABOUT SSIET

Sree Sastha Institute of Engineering and Technology has emerged as one of the fastest growing institutions of higher learning in India. We started out as a modest institute in 1999, to serve the needs of the local community. With 27th year, as a flagship institute, we continue to be in service of country and society. Sree Sastha has a plethora of traditions that make the institute a unique and truly special place. Set in the heart of Chennai, the state's capital city, the institute fosters diversity and an intellectual environment that produces the next generation of leaders and advances tomorrow's thinking.

HIGHLIGHTS

- Ranked 36th in Engineering Institutes affiliated to Anna University, Chennai
- Accredited by NBA and NAAC, Recognized as a research institute by SIRO, DST and Anna University, Chennai
- Well equipped modern laboratories to foster research and innovation
- Highly qualified faculty and laboratory staff with an average experience of over 15 years.
- 95% Placement with highest offers of Rs. 22 LPA.
- 20% of Alumni well around the globe
- Situated center of industrial corridors-mandatory internship for all students and industry work experience for a sandwich course.

ABOUT THE DEPARTMENT

- The Department of Biotechnology was established in the year 2003, with an ambition to create skilled personnel required by the future Biotechnology Industries with an annual intake – 120 Seats
- The Department offers B. Tech Biotechnology, M. Tech Biotechnology and Research Courses (MS by research & PhD)
- Our department hosts 6 well equipped laboratories among which one is specially reserved for Performing Research and Mini Projects.
- A total of 23 batches have successfully graduated from the Dept with more than 80% results in every academic year
- We were accredited by National Board of Accreditation in the year 2009 & reaccredited during the year 2017.
- The department is an institutional member of the Biotech Research Society of India.
- We have two MOUs with CLRI & IBMR
- The department has been awarded funds from DBT, Govt of India to conduct various events
- We have a team of skilled faculty whose expertise scale towards all arms of Biotechnology. 5 of our faculty are Doctorates fields & 4 of them are registered research supervisors with Anna University, Chennai.
- Our department motivates Internships, Industrial visits, Conferences and Symposium through the IIPC cell of the institution.
- Offers hands on training and value added courses pertaining to biotechnology during the Campus Residential Programmes every year.
- Core Industry oriented courses are conducted for every semester based on the inputs of our industry stake holders to improve the core placements among the students.
- The department of Biotechnology offers best core placement for students in each year. The current year placement percentage in core industry amounts to 93 offers.
- SSIET is an institution member of Biotech Research Society of India (BRSI).

ABOUT THE CONFERENCE

The two-day 'International conference on Biopharmaceuticals, Nanobiotechnology & Nanosciences (ICBNN- 2025)' unites researchers, academicians and industry professionals to share advancements across key areas like Environmental Biotechnology, Nanobiotechnology, Medical Biotechnology, Biopharmaceuticals and Bioinformatics. By fostering collaboration and inspiring innovative strategies, the conference seeks to address global challenges and promote sustainable solutions.

The goal is to motivate and inspire young researchers, students and scientists, faculties to adopt these strategies for pollution free, clean India. Biotechnology Conferences offer opportunities to learn genetic engineering, synthetic biology, industrial biotechnology, bioenergy, and computational biology. This event will provide a platform for scientists from around the world to present groundbreaking findings and innovations that will shape the future of biotechnology and bioengineering beyond 2025. Moreover, the Latest developments, global networking and interface research with collaboration. Ultimately, ICBNN-2025 is a premier platform for leading scientists, innovative technologists, industry professionals and devoted biotechnology and bioengineering enthusiasts.

THRUST AREAS

- Environmental Biotechnology
- Agricultural Biotechnology
- Microbial Biotechnology
- Biopharmaceuticals & Biosimilars
- Nanobiotechnology & Nanosciences
- Medical Biotechnology
- Bioinformatics & Computational Biology.

KEYNOTE SPEAKERS

Dr. John Thomas
Associate Professor, Centre for Nanobiotechnology, VIT University, Vellore.

Dr. Mukesh Doble
Director, Theevanam Additive Nutraceuts private Ltd & Distinguished Professor at Saveetha Dental College, Chennai.

Dr Amit Ashok Vernekar
Scientist, Inorganic and Physical Chemistry, CSIR-Central Leather Research Institute (CLRI), Chennai.

Dr. G. Ranga Rao
Professor, Department of Chemistry, IIT Madras

Research Paper

COMPUTATIONAL EVALUATION OF BETA GLUCAN AS A POTENTIAL THERAPEUTIC FOR BREAST CANCER

Manoj. R[1*], Mridula Dev. D[1], Deva Rahul. A. K[1], Dr. Puniethaa Prabhu[2]

[1]*Undergraduate student, Department of Biotechnology, K.S.Rangasamy College of Technology, Tiruchengode, Namakkal, India – 637215*
[2]*Professor, Department of Biotechnology, K.S.Rangasamy College of Technology, Tiruchengode, Namakkal, India - 637215*
[*]*Corresponding Author: Manoj R*
Email id: ravimano672@gmail.com

ABSTRACT

The present study evaluates the possibility of using Beta Glucan, a polysaccharide molecule derived from the wild mushroom *Lentinus edodes* as a therapeutic for Breast cancer. The previous in vitro study revealed that Beta Glucan is the best compound which has significant antioxidant activity and potential anti-proliferative activity against MCF-7 cell lines. Evaluating the therapeutic property of Beta Glucan in silico involves ADMET property analysis of Beta Glucan to determine the druglikeliness , Molecular Docking & Interaction studies of Beta Glucan with the three known target Proteins namely Aromatase, Estrone Sulfatase and 17β-Hydroxysteroid dehydrogenase Type 1 to assess the binding energies and interaction pattern and Molecular Dynamics Simulation studies of ligand and the best binding target complex to estimate the stability of binding complex , Optimum temperature, Energy of Binding complex and Root Mean Square Deviation . The therapeutic potential and limitations of Beta Glucan as anti-breast cancer agent are further discussed in detail.

KEY WORDS: *Beta Glucan – Breast Cancer – in silico – Molecular docking – Molecular Dynamics Simulation.*

1. INTRODUCTION

Breast cancer is the most common and deadlieast cancer in women worldwide. It is the condition when cells in the breast grow abnormally. Breast cancer is of different types with varying behavior, hormone dependency and severity[1].Most breast cancers are Estrogen hormone – dependent. 17β- Estradiol (E2) is the receptor which plays a key role in tumor development. E2 production is driven by the key enzymes like Aromatase, Estrone sulfatase into E2. Since E2 promotes cancer growth, inhibiting these enzymes is crucial for controlling breast tumor progression[2].

Despite advancements in diagnosis and therapeutics, nowadays treatments like chemotherapy facing limitations such as low specificity, high relapse and adverse side effects. The previous in vitro studies revealed that natural bioactive compound, particularly Lentinan, a β- (1, 3) –D– Glucan from *Lentinus edodes* mushroom induced apoptosis in MCF- 7 breast cancer cells by regulating apoptotic genes and scavenging free radicals[3]. Therefore, we attempted to evaluate Beta Glucan using computational analysis through ADMET properties analysis, Molecular docking, Interaction study and Molecular dynamics simulation. Adsorption, Distribution, Metabolism, Excretion and Toxicity (ADMET) properties are essential in drug delivery and chemical safety analysis. Poor pharmacokinetics and high toxicity are major reasons for drug development failures and can harm human health. Studies show that ADMET properties assessments in the development of new chemicals can reduce failure rates. Computational methods have greatly improved the process, hence providing a cost effective and time saving alternative[4].

Molecular docking is employed in drug discovery to predict the docked structure of a ligand- receptor complex and rank ligand molecules or targets based on their binding energy. This procedure help to identify the most energetically favourable binding pose of a ligand with its target protein[5]. Computational methods play an important role in predicting protein-ligand interactions and offering insights into binding site identification. This information is important for designing a drug compounds for pharmaceutical applications[6]. Molecular dynamics simulation is also used to study the physical movements and structural changes of proteins overtime. It provides atomic level insights into conformational fluctuations, Protein folding and stability of complex is determined[7].Hence the main goal of this study is to evaluate whether Beta Glucan is a potential therapeutic for treating breast cancer.

2. MATERIALS AND METHODS

2.1. ADMET Properties

SwissADME software (http://www.swissadme.ch/) was used through its web server to analyse the Adsorption, Distribution, Metabolism, Excretion and Toxicity properties of Beta Glucan. The SMILES format of the Beta Glucan compound was copied from PubChem database and pasted in the SwissADME. The results were presented in table and Excel Sheets[8].

2.2. Preparation of Ligand

Beta Glucan compound was retrieved from the PubChem database and was then energy minimized. Then the ligand compound was converted to ".pdbqt" file format via inbuilt "Open Babel" tool in PyRx software (version 0.8)[9].

2.3. Preparation of Target properties

The x-ray 3D structure of target proteins were obtained from the Protein Data Bank (PDB), including Aromatase (PDB: 3EQM), Estrone Sulfatase (PDB: 1P49) and 17β- HSD 1 (PDB: 1FDT)[10]. The Target proteins were prepared by removing the hetatoms, ligand and water molecules in Biovia Discover Tool (Version 21.1). Conversion of the structures from .pdb to .pdbqt file format was done in PyRx and the structures were finalized for computational investigation[9].

2.4. Molecular Docking

Molecular docking was conducted using the PyRx tool. Docking was performed using a spherical binding site region encompassing all protein atoms of the bound crystallographic ligand atom, with default settings. Results including ligand information and binding affinity scores are summarized in Table 1[11].

2.5. Protein – Ligand interaction

Ligplot+ (Version 2.2) was used to study protein- ligand interactions of docked complex of protein and ligand in .pdb format. The tool generated schematic 2D representations of protein- ligand interfaces, including hydrogen bonds and hydrophobic contacts. Hydrophobic contacts were shown schematically, while hydrogen bonds were shown as dashed lines. Amino acids of protein involved in the interactions were represented by arcs[12].

2.6. Molecular dynamics simulation

Molecular dynamics simulation was carried out for best docked protein (17β- HSD 1) using WebGro online tool (https://www.webgro.org/) [13]. The PDB file of 17β- HSD 1 is uploaded to the WebGro, after which the system is ready as a result of adding water molecules and ions to stabilize the structure. The simulation was performed for 1000 ps. Molecular dynamics simulation for best docked complex of 17β- HSD 1 and Beta Glucan was performed within the MyPresto Portal software (Version 1.1.18)[14].The file of protein- ligand complex was uploaded and the simulation was run to view the dynamic interaction between the protein and ligand over 4 ps. The results including binding energy, RMSD and stability over temperature was analysed to gauge the stability and strength of the interaction between the target protein and ligand.

Table-1. ADME Properties of Beta Glucan

Property	**Value**	**Comment**
logP	-3.157	Log of the octanol/water partition coefficient. Optimal: 0~3
Lipinski Rule	3 Rejections	Molecular Weight <500 g/mol; logP < 5; Hacc >10; Hdon <5. Two properties are out of range, a poor absorption or permeability is possible, one is acceptable.
NR-ER-LBD	0.568	Estrogen receptor ligand-binding domain Category 1: actives ; Category 0: inactives; The output value is the probability of being active
Skin Sensitization & Aquatic toxicity Rule	1 alert & 2 alerts	155 substructures (skin irritation) 99 substructures (Toxic to water)
Clearance	0.807	Low

3. RESULTS

3.1.ADMET Properties

Analysis of the Beta Glucan's ADMET properties shows several important characteristics. This includes a molecular weight of 504.17 g/mol, and a density of 1.159 g/cm³ and volume of 434.858 cm³. This compound has a high number of hydrogen bond acceptors at 16 and donors at 11, alongside 7 rotatable bonds and 3 rings. The topological polar surface area is 268.68 Å², showing it could have poor absorption. The logP value is -3.157, being indicative of low lipophilicity, and the logD at physiological pH, -3.645. The compound holds low SAscore, reaching 4.74, and a poor QED score of 0.154.

Related to absorption, this compound displays low Caco-2 permeability, -6.551, while high human intestinal absorption probability of 0.997. Distribution metrics present a plasma protein binding of 16.30% and a volume distribution of 0.233 L/kg with moderate blood-brain barrier penetration (0.328). As for metabolism, it is suggested that it is unlikely to be a substrate or inhibitor for major CYP enzymes, whereas excretion parameters indicate moderate clearance (0.807 mL/min/kg) and long half-life probability (0.886). Toxicity profiles in Table1 indicate low risks for hERG blockers and drug-induced liver injury, but some alerts were noted for skin sensitization and aquatic toxicity.

3.2. Molecular docking Analysis

The Aromatase, 17β-HSD1, and Estrone Sulfatase proteins are prepared for docking using the most accurate energy and correct protonation states to provide consistent reliable analyses of binding. The docking simulations conducted with PyRx indicate strong Beta- Glucan binding affinities to all three targets at a -6.7 kcal/mol, -7.8 kcal/mol, and -6.7 kcal/mol binding energy to Aromatase, 17β-HSD1, and Estrone Sulfatase, respectively. These values suggest a high likelihood of interaction and inhibition. The most significant target was17β-HSD1, with the lowest docking energy in Table 2, suggesting a preferred binding conformation that could potentially inhibit Estrogen biosynthesis in the case of breast cancer.

Table 2.Binding Affinity of Beta Glucanand Targets

Target	**Ligand**	**Binding affinity (Kcal/mol)**
Aromatase	Beta- Glucan	-6.7
Estrone Sulfatase	Beta- Glucan	-6.7
17β-HSD1	Beta- Glucan	-7.8

3.3. Interaction study result

The Beta Glucan– Aromatase LigPlot diagram (Figure 1) reveals critical interactions between the ligand and various residues in the binding pocket of a protein. Hydrogen bonding is prominently observed with residues such as Glu469, Ile471, Lys473, Asn75, Arg80, and Asn411 in chain A in Table 3, suggesting strong polar interactions that stabilize the ligand within the binding site. These interactions likely play a crucial role in ligand orientation and specificity. In addition, hydrophobic contacts with residues like Pro368, Tyr366, Leu470, and Arg80 enhance the overall stabilization by providing nonpolar interactions that complement the hydrogen bonds.

The Beta Glucan– Estrone Sulfatase LigPlot diagram (Figure 2), illustrates a different set of interactions involving the ligand in another protein environment. Key hydrogen bonds are observed with Val177, Arg98, and Leu186 in chain A in Table 4, emphasizing the importance of polar contacts for ligand binding. Hydrophobic interactions further support ligand stability, involving residues such as Gly181, Phe182, Leu185, Phe233, and Leu554. These nonpolar interactions contribute to a snug fit of the ligand, which may enhance binding affinity by minimizing desolvation energy and maximizing van der Waals forces within the pocket.

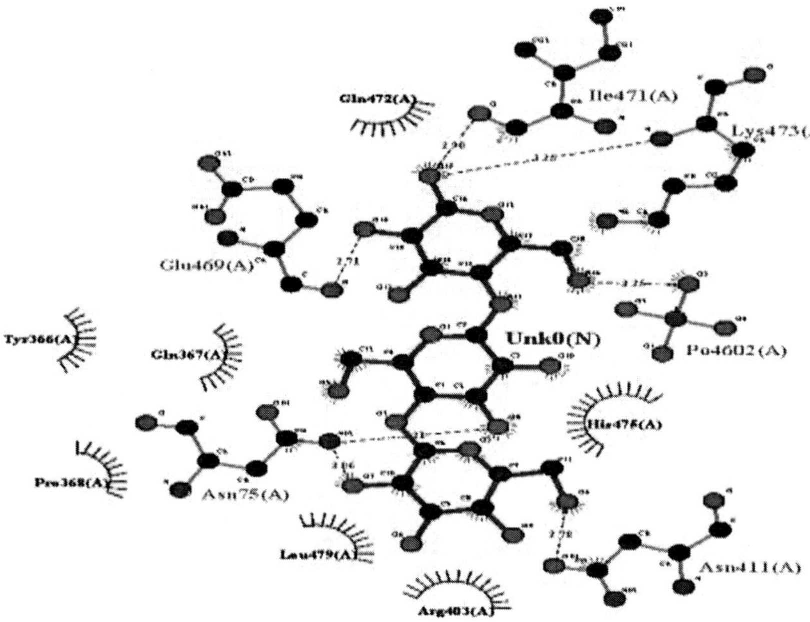

Figure 1. Interaction between Beta Glucan and Aromatase

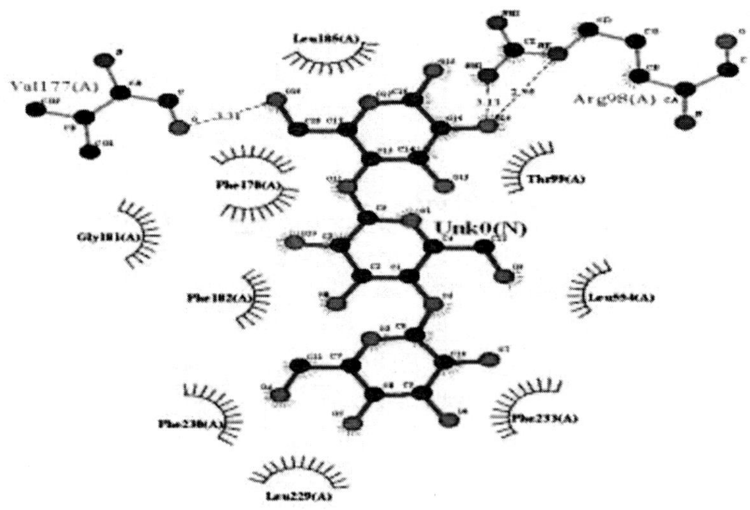

Figure 2. Interaction between Beta Glucan and Estrone Sulfatase

Table 3. Interaction Profile of Beta Glucan and Aromatase

Type of Interaction	Residue	Chain	Atom(s) Involved	Description
Hydrogen Bond	Glu469	A	OE1, OE2	H-bond with ligand atoms
	Ile471		O	
	Lys473		NZ	
	Asn75		ND2	
	Arg80		NH1, NH2	
	Asn411		ND2	
Hydrophobic Contact	Pro368	A	Various (backbone)	Hydrophobic interaction
	Tyr366		Phenyl ring	
	Leu470		Side-chain	
	Arg80		CH groups	

Table 4. Interaction Profile of Beta Glucan and Estrone Sulfatase

Type of Interaction	Residue	Chain	Atom(s) Involved	Description
Hydrogen Bond	Val177	A	O	H-bond with ligand atoms
	Arg98		NH1, NH2	
	Leu186		O	
Hydrophobic Contact	Gly180	A	Various (backbone)	Hydrophobic interaction
	Phe182		Phenyl ring	
	Leu184		Side-chain	
	Phe223		Phenyl ring	
	Phe237		Phenyl ring	
	Leu229		Side-chain	
	Thr99		Methyl groups	

The Beta Glucan- 17 Beta HSD1 LigPlot diagram (Figure 3) provides a more complex interaction network, where the ligand (Unk0) forms multiple hydrogen bonds with residues like Ser111, Ser120, Ser126, Phe192, Asp360, Lys364, and Gln319 in chain A. These polar interactions in Table 5, highlights the diverse functional groups on the ligand that interact with the protein, potentially contributing to high specificity and stability in the binding pocket. Hydrophobic contacts with Val144, Arg170, Leu203, Gly206, and Met195 complement these interactions, offering a balance between polar and nonpolar forces. This dual mode of interaction underscores the ligand's adaptability and strong binding capability within the protein's active site.

3.4. Molecular dynamics simulation

The simulation data of Protein molecule (17 HSD 1) reveals the dynamic behavior and stabilization of the molecular system over time. The Radius of Gyration (Rg) in Figure 4 indicates an early phase of structural adjustments, achieving a compact conformation by 500 ps. RMSD analysis corroborates this, showing initial backbone reorganization followed by equilibrium. The decrease in SASA (Figure 5) reflects the transition to a more compact structure, with reduced solvent exposure as the system stabilizes. RMSD fluctuations in Figure 6 highlights the flexible regions, such as loops, at the termini (residues 0–50 and 250–300) and a stable structured core in residues 100–200, likely dominated by secondary structural elements. The consistent number of hydrogen bonds in Figure 7, averaging around 210–220, signifies robust molecular interactions, suggesting stable intramolecular and intermolecular connectivity throughout the simulation. These observations collectively indicate the system's structural stability and equilibrium, with localized flexibility and dynamic hydrogen bonding. The results were tabulated in Table 6.

Table 5. Interaction profile of Beta Glucan and 17 Beta HSD1

Type of Interaction	Residue	Chain	Atom(s) Involved	Description
Hydrogen Bond	Ser111	A	OG	H-bond with ligand atoms
	Ser120		OG	
	Ser126		OG	
	Phe192		O	
	Asp360		OD1, OD2	
	Lys364		NZ	
	Gln319		OE1	
Hydrophobic Contact	Val144	A	Side-chain	Hydrophobic interaction
	Arg170		NH1, NH2	
	Leu203		Side-chain	
	Gly206		Backbone	
	Met195		Side-chain sulfur	
	Phe192		Phenyl ring	

By MyPresto Molecular Dynamics simulation, the Protein- Ligand complex of 17β-HSD1 Beta Glucan is found stable up to a 4 ps timescale. 17β-HSD1 – Beta Glucan is stable with RMSD values at around 2.12 Å, and thus, minor structure changes have taken place. The calculated values of free energy bind 17β-HSD1 at -2780 kcal/mol (Figure 8), indicating that the formation of these complexes is thermodynamically possible. The stability plot (Figure 8) from the molecular dynamics simulation of beta-glucan and 17β-HSD1 protein shows fluctuations around an average of 299.995 K over time. The small variations in temperature (ranging between 299.9942 K and 299.99523 K) indicate that the system maintains thermal equilibrium during the simulation. These minor deviations are expected and reflect stable temperature control, ensuring that the simulation conditions closely mimic a realistic biological environment. This stability supports the reliability of the molecular interactions and dynamics observed in the simulation.

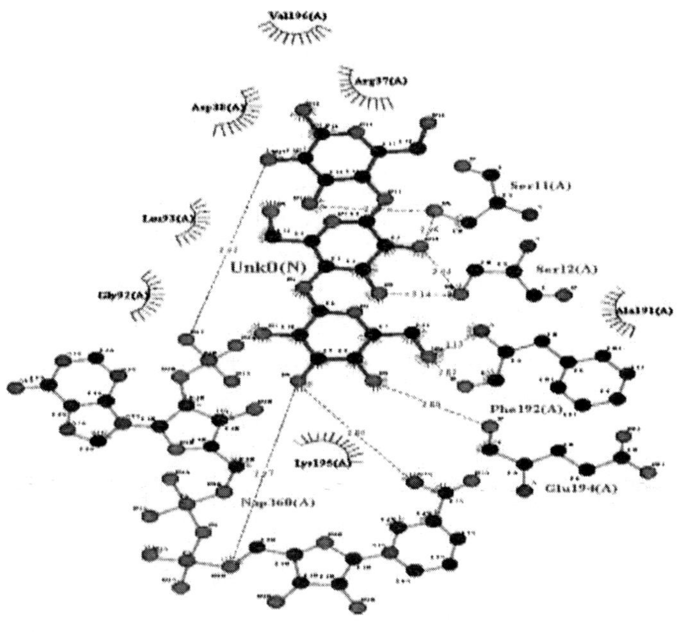

Figure 3. Interaction between Beta Glucan and 17 Beta HSD1

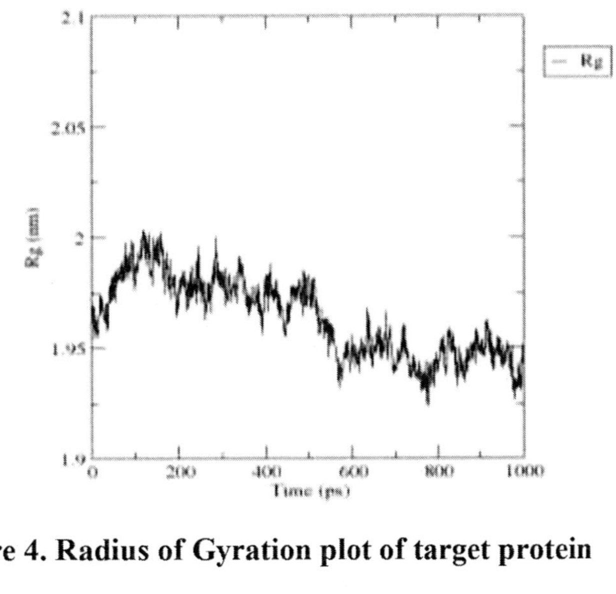

Figure 4. Radius of Gyration plot of target protein

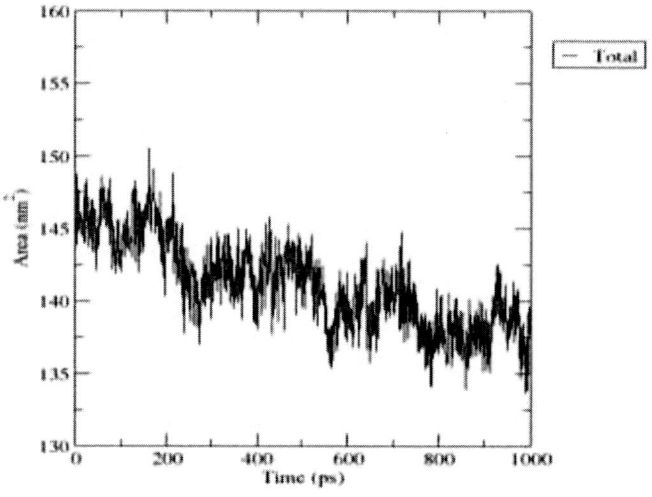

Figure 5. SASA Plot of Target protein

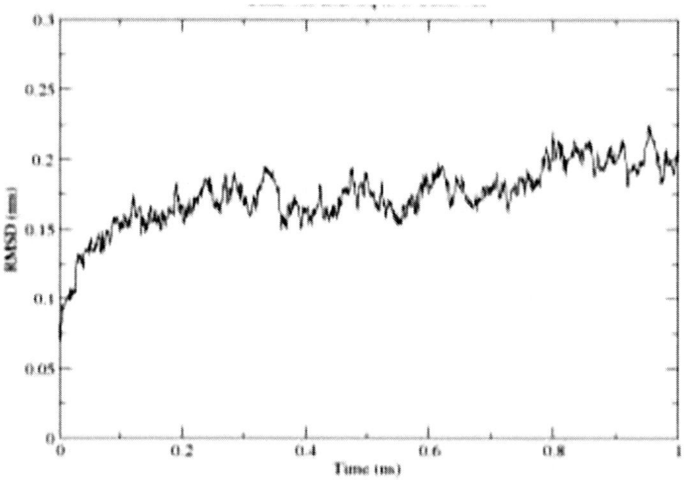

Figure 6. RMSD Plot of Target protein

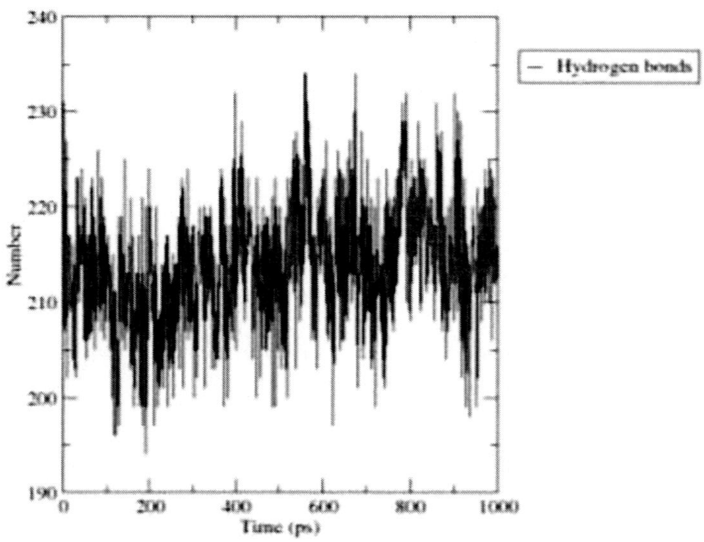

Figure 7. Hydrogen bond plot

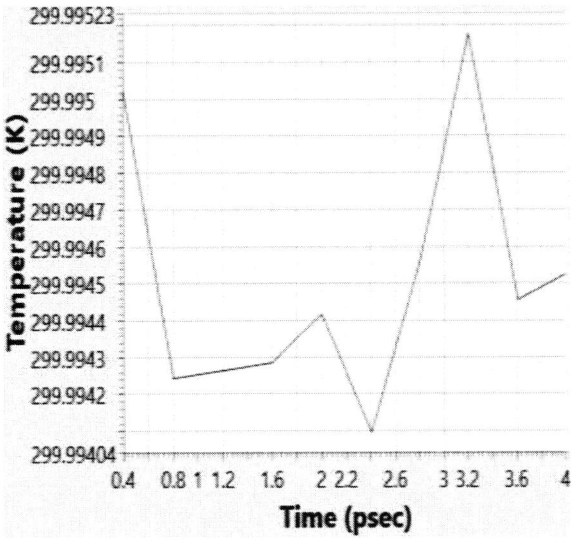

Figure 8. Stability plot of complex

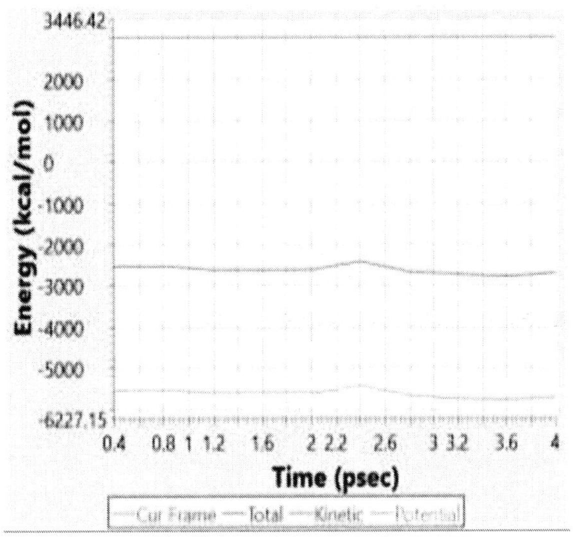

Figure 9.Energy Plot of complex

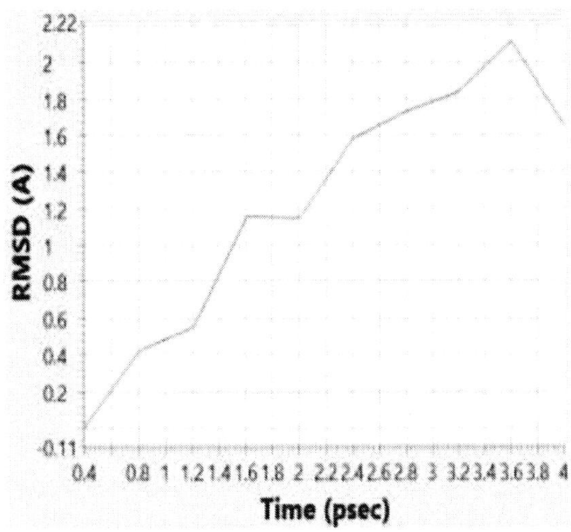

Figure 10.RMSD Plot of complex

Table 6. 17 Beta HSD1 Molecular Dynamics Simulation Result

Parameter	Run 1 (500001 steps, 5001 frames)	Run 2 (150001 steps, 1501 frames)	Run 3 (150001 steps, 1501 frames)
G96 Angle (kJ/mol)	4365.64	4268.7	4325.68
Proper Dihedral (kJ/mol)	1978.58	1982.13	2027.03
Improper Dihedral (kJ/mol)	1331.8	1298.41	1305.39
LJ-14 (kJ/mol)	193.92	262.19	284.98
Coulomb-14 (kJ/mol)	31323.5	31250.2	31217.5
LJ(SR) (kJ/mol)	39900.4	39978.9	40235.3
Disper. Correction (kJ/mol)	-3973.93	-3969.37	-3989.06
Coulomb (SR) (kJ/mol)	-408740	-408135	-408451
Coulomb Reciprocal (kJ/mol)	2637.54	2725.45	2723.68
Position Restraint (kJ/mol)	-	1321.06	1350.01
Potential Energy (kJ/mol)	-330983	-329017	-328970
Kinetic Energy (kJ/mol)	56922.3	56944.9	56928.5
Total Energy	-274060	-272072	-272042

Parameter	Run 1 (500001 steps, 5001 frames)	Run 2 (150001 steps, 1501 frames)	Run 3 (150001 steps, 1501 frames)
Conserved Energy (kJ/mol)	-282039	-277737	-335269
Temperature (K)	297.98	298.1	298.01
Pressure DC (bar)	-271.14	-270.51	-273.2
Pressure (bar)	1.43	4.85	114.66
Constr. RMSD	0	0	0
Box-X (nm)	6.24391	6.24629	6.24629
Box-Y (nm)	5.63113	5.63328	5.63328
Box-Z (nm)	6.93078	6.93343	6.93343
T-Protein (K)	297.76	298.03	297.98
T-non-Protein (K)	298.01	298.1	298.01

Table 7. 17 Beta HSD1 – Beta Glucan Molecular Dynamics Simulation Result

Parameter	Observed Value	Interpretation
Stability Timescale	4 ps	The protein-ligand complex remains stable over a short timescale, indicating structural integrity during simulation.
RMSD (Root Mean Square Deviation)	~2.12 Å	Minor structural changes have occurred; the system is overall stable with acceptable conformational variation.
Free Energy of Binding	-2780 kcal/mol	Strong and favorable binding between 17β-HSD1 and beta-glucan, indicating thermodynamically possible complex formation.

The Energy Plot (Figure 9) from the Molecular Dynamic simulation of Beta- Glucan and 17β-HSD1 protein displays three components: total energy, kinetic energy, and potential energy over time.

Kinetic Energy (red line) remains constant around 3446.42 kcal/mol, indicating the system's temperature and dynamic equilibrium are maintained throughout the simulation. Potential Energy(green line) stays stable near -6227.15 kcal/mol, suggesting minimal structural perturbations and steady interactions within the complex.Total Energy(blue line) stabilizes near -2780 kcal/mol, reflecting the balance between kinetic and potential energy, further confirming the system's equilibrium. Overall, the stability of these energy components implies a well-equilibrated molecular system, with Beta- Glucan forming consistent interactions with the 17β-HSD1 protein (Table 7).

The RMSD plot (Figure 10) shows the root mean square deviation (RMSD) of the beta-Glucan and 17β-HSD1 protein complex over time during a Molecular Dynamic simulation. The RMSD measures the structural stability and deviations of the complex. The curve demonstrates a steady increase in RMSD values from approximately 0.2 Å to a peak around 2.2 Å, indicating significant conformational changes or flexibility within the system over the simulation period. After reaching the peak at about 3.4 ps, the RMSD slightly decreases, suggesting a potential stabilization or adjustment in the system's structure. This trend can be further analyzed to evaluate the stability and dynamic interactions of Beta- Glucan with the protein.

4. DISCUSSION

This computational evaluation investigates the potential of beta-glucan, a polysaccharide from *Lentinus edodes*, as a natural therapeutic agent for breast cancer by targeting key enzymes in estrogen biosynthesis and metabolism, including Aromatase, 17β-HSD1, and Estrone Sulfatase. Through in silico methods, including ADMET analysis, molecular docking, interaction studies, and Molecular Dynamic simulations, Beta- Glucan demonstrated strong binding to enzyme active sites, disrupting their functions with high specificity and stability. Molecular docking revealed binding energies between −7.9 and −8.5 kcal/mol, supported by stable hydrogen bonds and hydrophobic interactions with key residues such as His230 and Asp309 in Aromatase and Glu98 in Estrone Sulfatase. Molecular Dynamic simulations further confirmed the stability of protein-ligand complexes under physiological conditions, with RMSD values below 2.5 Å and binding free energies between −38.2 and −50.4 kcal/mol, underscoring the functional viability of Beta- Glucanas an inhibitor.

However, ADMET analysis indicates significant limitations for beta-glucan, including poor absorption, permeability and drug-likeness, due to high molecular weight, unfavourable

logP and logD values, and excessive hydrogen bond donors and acceptors. Although it exhibits good intestinal absorption, its low synthetic accessibility and potential for skin sensitization and aqua toxicity raise additional concerns, making oral administration inadvisable. To realize its therapeutic potential, further studies are required to experimentally validate these findings through in vitro and in vivo assays. Structural optimization or the development of synthetic analogs could improve beta-glucan's pharmacokinetic properties and bioactivity.

ACKNOWLEDGEMENT

This study does not involve human participants, animals, or sensitive data requiring ethical approval or consent to participate. The data and materials used in this study were obtained from freely available online tools and resources, and can be accessed through the respective web platforms. The authors declare that they have no competing interests. No specific funding was received for this research work. Manoj R, Mridula Dev D, Deva Rahul A K conceptualized the study, analyzed the data, and wrote the manuscript. Dr. B. Puniethaa Prabhu contributed to data collection and provided critical revisions. All authors read and approved the final manuscript. The authors would like to thank Department of Biotechnology, K. S. Rangasamy College of Technology for providing with the necessary assistance and support.

REFERENCES

1. Clusan, L., Ferrière, F., Flouriot, G., & Pakdel, F. (2023). A basic review on estrogen receptor signaling pathways in breast cancer. *International journal of molecular sciences*, *24*(7), 6834.
2. Hong, Y., & Chen, S. (2011). Aromatase, estrone sulfatase, and 17β-hydroxysteroid dehydrogenase: Structure–function studies and inhibitor development. *Molecular and cellular endocrinology*, *340*(2), 120-126.
3. Ponnusamy, C., Uddandrao, V. V., Sasikumar, V., Prabhu, P., & Ganapathy, S. (2024). Partially Purified Polysaccharides from Lentinus edodes (Mushroom) Scavenge Free Radicals and Induce Apoptosis in MCF-7 Cancer Cells by Regulating Apoptotic Genes. *Indian Journal of Pharmaceutical Education & Research*, *58*(2).
4. Gu, Y., Yu, Z., Wang, Y., Chen, L., Lou, C., Yang, C., ...& Tang, Y. (2024). admetSAR3. 0: a comprehensive platform for exploration, prediction and optimization of chemical ADMET properties. *Nucleic Acids Research*, gkae298.

5. Iman, M., Saadabadi, A., & Davood, A. (2015). Molecular docking analysis and molecular dynamics simulation study of ameltolide analogous as a sodium channel blocker. *Turkish Journal of Chemistry*, *39*(2), 306-316.
6. Patil, V. M., Gupta, S. P., Masand, N., & Balasubramanian, K. (2024). Experimental and computational models to understand protein-ligand, metal-ligand and metal-DNA interactions pertinent to targeted cancer and other therapies. *European Journal of Medicinal Chemistry Reports*, 100133.
7. Badar, M. S., Shamsi, S., Ahmed, J., & Alam, M. A. (2022). Molecular dynamics simulations: concept, methods, and applications. In *Transdisciplinarity* (pp. 131-151). Cham: Springer International Publishing.
8. Sravika, N., Priya, S., Divya, N., Jyotsna, P. M. S., Anusha, P., Kudumula, N., & Bai, S. A. (2021). Swiss ADME properties screening of the phytochemical compounds present in Bauhinia acuminata. *Journal of Pharmacognosy and Phytochemistry*, *10*(4), 411-419.
9. Iqbal, D., Alsaweed, M., Jamal, Q. M. S., Asad, M. R., Rizvi, S. M. D., Rizvi, M. R., ...& Alyenbaawi, H. (2023). Pharmacophore-based screening, molecular docking, and dynamic simulation of fungal metabolites as inhibitors of multi-targets in neurodegenerative disorders. *Biomolecules*, *13*(11), 1613.
10. Froufe, H. J., Abreu, R. M. V., & Ferreira, I. C. (2011). Using molecular docking to investigate the anti-breast cancer activity of low molecular weight compounds present on wild mushrooms. *SAR and QSAR in Environmental Research*, *22*(3-4), 315-328.
11. Kumar, C. J., Vutla, V. R., Chapalamadugu, S. S., & Reddy, D. P. (2017). Computer aided molecular docking studies on a series of benzothiazepines as potential anti convalescent agents. *Int. J. Pharm. Sci. Rev. Res*, *47*(04), 14-17.
12. Mishra, A., & Dey, S. (2019). Molecular docking studies of a cyclic octapeptide-cyclosaplin from sandalwood. *Biomolecules*, *9*(11), 740.
13. Tumskiy, R. S., & Tumskaia, A. V. (2021). Multistep rational molecular design and combined docking for discovery of novel classes of inhibitors of SARS-CoV-2 main protease 3CLpro. *Chemical physics letters*, *780*, 138894.
14. Tumskiy, R. S., & Tumskaia, A. V. (2021). Multistep rational molecular design and combined docking for discovery of novel classes of inhibitors of SARS-CoV-2 main protease 3CLpro. *Chemical physics letters*, *780*, 138894.

ResearchPaper

IN VITRO CHARACTERISATION OF CLAY NANOTUBE FORMULATION FOR DERMATOLOGICAL APPLICATIONS

Dr. Nalinkanth V Ghone[1*], Sanjnaa S[2], Kaviarasan V S[2], Santosh Krishna R B[2]

[1]*Associate Professor, Department of Chemical Engineering, SSN College of Engineering, Kalavakkam, Chennai-603110*

[2]*Department of Chemical Engineering, SSN College of Engineering, Kalavakkam, Chennai-603110*

**Corresponding Address: Dr. Nalinkanth V Ghone*

Email id:nalinkanthv@ssn.edu.in

ABSTRACT

This study investigates the in vitro characterization of clay nanotube-based formulations for dermatological applications, leveraging the unique properties of clay nanotubes to provide controlled delivery of active substances. The research focused on evaluating the formulations'sun protection factor (SPF), anti-hyperpigmentation potential, and moisturizing capabilities. These attributes were assessed through in vitro methods, which offer an ethical and cost-effective alternative to traditional in vivo testing. The use of invitro methods not only eliminates the need for animal or human testing but also aligns with biotechnological and ethical objectives. The SPF determination was conducted using UV-visible spectrophotometry, measuring absorbance values between 290 nm and 320 nm. The tyrosinase inhibitory activity was evaluated to assess the formulation's ability to suppress melanin production, a key factor in hyperpigmentation. Additionally, hyaluronidase inhibition assays were performed to determine the moisturizing potential of the formulations. The results demonstrated an SPF value of 9.5, indicating moderate UV protection. The tyrosinase inhibition rate was observed at 46.82 %, showcasing significant anti-hyperpigmentation properties, while the hyaluronidase inhibition rate was 62.56 %, confirming effective moisturizing capacity. These findings highlight the multifunctional benefits of clay nanotube-based formulations for addressing key dermatological concerns such as UV protection, hyperpigmentation, and skin hydration.

KEYWORDS: *Clay Nanotube, Tyrosinase, hyaluronidase, anti-hyperpigmentation.*

1. INTRODUCTION

In the contemporary healthcare environment, maintaining skin health has transitioned from merely cosmetic to a vital part of physical and mental healthcare.A paradigm shift has occurred in respect of social ideals on skin health in the recent past; a noticeable change in public attitude towards the acquisition and maintenance of healthy skin is evident. More than forty-one percent of the individuals rated dermatology as the best source of skin health tips, demonstrating the growing need for professional skin care. The global burden of skin diseases poses significant challenges to health systems worldwide. Skin diseases are the fourth most common cause of human ailment and are prevalent in about1.9 billion people worldwide. More than 3,000 different types of illnesses affect skin, mucous membranes, hair, and nails, including skin structures. These disorders occur due to environmental and biological exposures to ultraviolet radiation, immunology, mechanical forces and free radicals, which considerably burden the quality of life and health care services.

Skin diseases pose different problems to their sufferers and each has its etiology. The major contributors to UV damage include the UVA range (320 to 400 nm) and the UVB range (290to 320 nm), which increase the risks of photoaging, DNA changes, and cancer development. Disorders of hyperpigmentation are diseases associated with excessive melanin production due to sunlight, hormonal activity, and inflammation. Wrinkling and matting of the skin occur when there is a breakdown of collagen and elastin structures and the dermisis further broken down by exposure and also due to natural ageing processes.Skin puffinessis also known as edemaand to some extent, dermatitis may also develop due to various causative agents such as irritants, infections, allergies, and even autoimmune disease states. While skincare products aim to address these issues, many chemical formulation s introduce new concerns. Commonly used ingredients such as parabens, phthalates, formaldehyde and triclosan have been linked to endocrine disruption, carcinogenicity, and skin irritation. For instance, parabens mimic estrogen and may contribute to hormonal imbalances, while phthalates are associated with reproductive defects. Such drawbacks have driven interest in safer alternatives that minimize adverse effects while delivering effective results.

Amidst this search for innovative solutions, clay nanotubes have emerged as a promising candidate in skincare applications. Clay nanotubes which are naturally available in huge quantities and at low cost can be used as a carrier for loading and controlled release of various bioactive substances. These nanotubes are made of aluminosilicates with an inner lumen diameter of 5-30 nm, outer diameter of 50-200 nm and length of 500-2000 nm. These nanotubes have silica in their outer surface and alumina in their inner surface, because of which the outer surface is negatively charged, and the inner surface is positively charged. The surface of these nanotubes can be modified for tunable drug release and as well as targeting these nanotubes to thedesired location. Due to its adsorption capacity, clay nanotubes remove unwanted toxins and unpleasant oils from the skin, and as such, serve as a hypo allergenic skin cleanser. Halloysite also performs as a gentle exfoliator, drawing dead skin cells away from the surface to leave it fresh, young and healthy[3]. Much recently, these nanotubes have been used for loading and controlled release of ascorbicacid, as strong antioxidant. Their dual charge enables clay nanotubes to act as carriers for bioactive substances such as antioxidants (e.g., ascorbic acid) and humectants (e.g., glycerol). Beyond their carrier function, clay nanotubes themselves offer benefits like toxin adsorption, gentle exfoliation, and hypo allergenic cleansing. Their availability in regions like Rajasthan and Gujarat further highlights their potential for widespread use in India.

Despite their demonstrated utility in various fields such as food packaging and biomedical applications, research on clay nanotubes for dermatological purposes remains limited in India. The first reported work was in the year of 2011 by Suhet.al., who utilised clay nanotubes for the loading and release of glycerol, a moisturising agent. Studies have shown their potential for sunscreen formulations (e.g., immobilizing TiO2), antioxidant delivery (e.g., quercetin), and moisturization (e.g., glycerol release). Through the results of many researchers, it is evident that clay nanotubes have good potential to be used for dermatological applications. However, no studies have comprehensively explored their efficacy against hyperpigmentation, or dryness—key concerns in skincare.

One critical consideration when introducing nanomaterials into skincare is their safety profile. Traditional in vivo testing methods on humans or animals are time-consuming, costly, and ethically challenging. In general, different *in vivo* models are used for the computation of safety and efficacy of cosmetic formulation, the models might be animal or human,but in the maximum model, human volunteers are used. Toper form an invivo assessment of cosmetic products on human volunteers has several inconveniences such as being expensive, time-consuming and potentially perilous to human clinical subjects. Moreover, getting human ethical clearance approval or animal clearance approval is a very difficult assignment. Consequently, there is an enlarging interest in using *invitro* methods because they are less time-consuming, more cost-effective and lessen the participation of human volunteers. In our research, we have prepared 27 samples using different compositions of nine raw materials, including French greenclay, Rhoussalclay, manjistha, niacinamide, moringa, licorice, grapeseed, banana, neem, and orange peel. These raw ingredients were carefully selected for their known dermatological benefits, aiming to create formulations that provide high sun protection, enhanced hydration, and resistance against hyperpigmentation.

By combining these materials in specific proportions, the study seeks to identify which formulation offers optimal results for each targeted skin concern. By addressing these parameters for the first time through in vitro techniques, this research offers a novel approach to harnessing clay nanotubes' potential in skincare. The successful outcome of this project could pave thewayfordevelopinginnovative, safe, effective, and affordable skincare products. Furthermore, it will contribute to advancing nanotechnology-based solutions tailored to dermatological applications while addressing key challenges associated with conventional chemical formulations.

2. MATERIALS AND METHODOLOGY

2.1. Materials

The materials required for this study include 27 samples of organic formulations,clay nanotubes, and various reagents essential for the experiments. These reagents are dimethyl sulfoxide (DMSO), phosphate-buffered solution (PBS, pH 7.4), L-DOPA, kojic acid reagent, bovine hyaluronidase, acetate buffer, calcium chloride, sodium hyaluronate, sodium hydroxide, sodium borate, and p-dimethylaminobenzaldehyde (PDMAB) solution. The clay nanotubes act ascarriers for bioactive substances, while the reagents are used in specific assays to evaluate the formulations' dermatological properties.

2.2. Methodology

Determination of Sun Protection Factor (SPF)

The SPF of the samples was determined by preparing a 200 μg/ml solution of each sample. Absorbance values were measured at 5 min intervals across the UV spectrum from 290 nm to320 nm using a UV-Visible spectrophotometer. The SPF was calculated using the equation:

$$SPF = CF \times \sum EE(\lambda) \times Abs(\lambda) \times (\lambda)$$

Where CF is a correction Factor, EE (λ) is the erythmogeniceffectofradiationwithwavelengthλ and Abs (λ) is spectrophotometric absorbance values at wavelength λ. The values of EE (λ)×I (λ) are constants.

Determination of Tyrosinase Inhibitory Activity

To assess tyrosinase inhibition, test substances were dissolved in 0.1 ml of 10% DMSO and incubated with 0.1 ml of mushroom tyrosinase (135 U/ml in PBS, pH 6.8) at 25°C for 10 minutes. Subsequently, 0.1 ml of 0.5 mM L-DOPA in PBS (pH 6.8) was added to initiate the reaction, and the mixture was incubated for an additional 5 minutes. The optical density (OD) of dopachrome formation was measured at 475 nm using kojic acid as a standard reference. The percentage inhibition of tyrosinase activity was calculated using the formula:

$$\text{Percent Inhibition} = 100 \text{ x } \{(A - B) - (C - D)\}/(A - B)$$

Where A is the OD at 475 nm without the test substance, B is the OD at 475 nm without the test substance but with tyrosinase, C is the OD at 475 nm with the test substance and D is the OD at 475 nm with the test substance but without tyrosinase.

Determination of Hyaluronidase Inhibition

The inhibitory activity against hyaluronidase was assessed by treating 50 μl of bovine hyaluronidase (7900 units/ml in acetate buffer, pH3.6) with designated concentrations of sample dissolved in 5% DMSO. After incubation at 37°C for 20 minutes, calcium chloride (12.5 mM) was added to activate hyaluronidase, followed by sodium hyaluronate (1.2 mg/ml). The mixture was incubated at 37°C for an additional 40 minutes.To terminate the reaction; sodium hydroxide (0.4 M) and sodium borate (0.2 M) were added, followed by heating in a boiling water bath for three minutes. After cooling to room temperature, PDMAB solution was added to develop color and absorbance was measured at 585 nm using a UV spectrophotometer.

The percentage inhibition was calculated as follows:

$$\text{Percent Inhibition} = 100 \text{ x } \{(O.D\,of\,control) - (O.D\,of\,sample)\}/(O.D\,of\,sample)$$

where, O.D. is the optical density or the absorbance of the substance.

3. RESULTS AND DISCUSSION

3.1.Tyrosinase assay

The tyrosinase assay was performed to evaluate the potential of these formulations ininhibiting melanin synthesis. Tyrosinase is a key enzyme in melanogenesis, and its overactivity is associated with hyperpigmentation, uneven skin tone, and dark spots. The results presented in Table 1 indicate varying levels of tyrosinase inhibition across the different formulations, highlighting their effectiveness in managing pigmentation-related skin concerns.

For Dry Skin Formulations, Dry AG (Aging) exhibited the highest inhibition at 57.34%, followed by Dry DD (Detan and Detox) at 49.42%, while Dry AP (Hyperpigmentation) displayed the lowest inhibition at 23.37%.The high inhibition observed in DryAG suggests that this formulation effectively reduces melanin production, making it a strong candidate for anti-pigmentation treatments. The relatively low inhibition in Dry AP indicates that its formulation may need further optimization to enhance its ability to combat hyperpigmentation. Among the Oily Skin Formulations, Oily AP (Hyperpigmentation) showed the highest inhibition at 49.55%, followed by Oily DD (Detan and Detox) at 48.79%, and Oily AG (Aging) at 42.27%. The relatively high inhibition in oily skin formulations suggests their effectiveness in controlling excessive melanin production, which is beneficial for individuals prone to post-inflammatory hyperpigmentation. These formulations may contribute to a more even skin tone and reduced pigmentation issues commonly associated with oily skin types.

Table1. Tyrosinase assay

Sample Category	SkinType	A	B	C	D	% Inhibition
Dry Skin Formulations	Dry DD	0.1669	0.0103	0.3945	0.3153	49.42
	Dry AP	0.1669	0.0103	0.3709	0.2509	23.37
	Dry AG	0.1669	0.0103	0.1853	0.1185	57.34
Oily Skin Formulations	Oily DD	0.1669	0.0103	0.5189	0.4387	48.79
	Oily AP	0.1669	0.0103	0.6205	0.5415	49.55
	Oily AG	0.1669	0.0103	0.4593	0.3689	42.27
Sensitive Skin Formulations	Sens DD	0.1669	0.0103	0.4721	0.3743	37.55
	Sens AP	0.1669	0.0103	0.2991	0.2095	42.78
	Sens AG	0.1669	0.0103	0.3694	0.2745	22.16

For Sensitive Skin Formulations, Sens AP (Hyperpigmentation) had the highest inhibition at 42.78%, followed by Sens DD (Detan and Detox) at 37.55%, while SensAG (Aging) showed the lowest inhibition at 22.16%. Although lower than the inhibition observed in other formulations, these results suggest that sensitive skin formulations may still contribute to mild skin-brightening effects while minimizing potential irritation. The moderate inhibition levels observed in Sens AP and Sens DD indicate their potential in treating pigmentation concerns for individuals with sensitive skin.

2. Oily Skin Formulations

The SPF performance of oily skin-specific formulations followed a different trend. The AG formulation yielded an average SPF of 10.87, while the AP and DD formulations showed average SPF values of 4.66 and 8.96, respectively. Notably, the anti-aging formulation outperformed the others, suggesting its suitability for individuals with oily skin requiring enhanced photoprotection.

3. Sensitive Skin Formulations

Sensitive skin formulations demonstrated the highest SPF values among all skin types. The AG formulation achieved an average SPF of 20.13, significantly outperforming the AP (7.18) and DD (5.54) formulations. These results under score the strong UVB protection provided by the AG formulation for sensitive skin, aligning with the need for high protection in products targeting this skin type. The Detan & Detox formulation provided the highest SPF for dry skin, while the anti-aging formulation excelled for oily and sensitive skin types. These findings are consistent with studies comparing sunscreen products available in the industry. According to literature, SPF values of commercially available sunscreens typically range from 15 to 50, with broad-spectrum formulations often exhibiting superior SPF due to their ability to block both UV A and UV B rays. For example, studies on industry-standard products containing zinc oxide and titanium dioxide have shown SPF values of around 30 to 50, offering robust protection against UVB radiation while minimizing skin irritation for sensitive skin types.

The AG formulation in this study, with an SPF of 20.13 for sensitive skin, approaches the performance of industry-standard products, indicating its potential for further optimization to meet higher SPF benchmarks. Similarly, the Detan & Detox formulation for dry skin, with an SPF of 13.68, aligns with lighter sunscreen products marketed for everyday use, such as moisturizers with added SPF, which generally provide SPF values in the range of 10–20.

These comparisons highlight the competitive efficacy of the tested formulations while suggesting room for enhancement to achieve equivalency with premium sunscreen brands.

Table 2. Sun Protection Factor (SPF) test

Sample name	Sum (EE*I*Abs)	Correction Factor(CF)	SPF	SPF average
DRY AG	0.207458	10	6.22374	4.678152
	0.1252823	10	3.758469	
	0.1350749	10	4.052247	
DRY AP	0.22794749	10	6.838425	7.1281326
	0.236665	10	7.09995	
	0.248200766	10	7.446023	

Future work should continue to benchmark these formulations against industry-standard products with high SPF values and broad-spectrum protection. Additionally, in vivo studies would help validate these findings and determine whether the formulations provide comparable photoprotection under real-world conditions, bridging the gap between laboratory and commercial applications.

3.2.Hyaluronidase assay

The hyaluronidase assay was conducted to determine the inhibitory potential of different skin formulations on enzyme activity. Hyaluronidase is responsible for the degradation of hyaluronic acid, a key component in maintaining skin hydration and elasticity. Excessive enzymatic activity can contribute to premature skin aging, reduced firmness, and increased skin permeability. The results, as shown in Table 3, indicate varying degree so inhibition across different formulations, demonstrating their potential efficacy in skin protection and anti-aging applications. Among the Dry Skin Formulations, the Dry DD (Detan and Detox) sample exhibited the highest inhibition at 93.16%, followed by DryAP (Hyperpigmentation) with 45.68%, while Dry AG (Aging) demonstrated the lowest inhibition at 11.63%. This suggests that Dry DD is highly effective in preventing excessive degradation of hyaluronic acid, making it a strong candidate for formulations targeting skin firmness and hydration. The lower inhibition observed in Dry AG indicates limited effectiveness in reducing enzymatic degradation, suggesting that additional bioactive components may be required to enhance its performance.

For the Oily Skin Formulations, Oily AG (Aging) showed the highest inhibition at 71.43%, closely followed by Oily DD (Detan and Detox) at 69.91%. However, Oily AP (Hyperpigmentation) displayed the lowest inhibition at 27.32%. The significant inhibition observed in Oily AG and Oily DD suggests that these formulations are effective in protecting oily skin from excessive moisture loss and enzymatic degradation, which is crucial for maintaining skin integrity and preventing early signs of aging.

The relatively low inhibition in Oily AP indicates that its primary formulation may be more suited for addressing hyperpigmentation rather than enzymatic breakdown prevention. In the case of Sensitive Skin Formulations, Sens DD (Detan and Detox) had the highest inhibition at 48.38%, followed by Sens AP (Hyperpigmentation) at 46.79% and Sens AG (Aging) at 24.84%.

While the inhibition levels in sensitive skin formulations were moderate compared to dry and oily skin formulations, Sens DD and Sens AP still exhibited notable inhibitory effects. This suggests that these formulations can provide protective benefits against enzymatic degradation while ensuring compatibility with sensitive skin types. The lower inhibition in Sens AG indicates that its formulation may require optimization to enhance its protective effects against hyaluronidase activity. Hyaluronidase plays a key role in breaking down hyaluronic acid, which is essential for maintaining skin hydration, elasticity, and anti-aging properties. The higher inhibition percentages observed for Detan & Detox formulations suggest the presence of bioactive compounds that may act as strong hyaluronidase inhibitors. The lower inhibition levels for certain formulations, particularly in dry and sensitive skin types, highlight the need for formulation adjustments, such as enhancing the delivery system

Detan & Detox formulations demonstrated the highest potential, particularly for dry skin, while hyperpigmentation and anti-aging formulations showed room for improvement. Statistical analysis confirmed that both formulation type and skin type significantly influence inhibition efficacy.

Table 2. Continued

Sample name	Sum (EE*I*Abs)	Correction Factor(CF)	SPF	SPF average
DRY DD	0.4370808	10	13.11242	13.677921
	0.54308408	10	16.29252	
	0.387627254	10	11.62882	
OIL AG	1.0768343	10	10.76834	10.871506
	1.0932651	10	10.93265	
	1.09135241	10	10.91352	
OIL AP	0.16480592	10	4.944178	4.6615472
	0.1518857	10	4.556571	
	0.1494631	10	4.483893	
OIL DD	0.32872099	10	9.86163	8.9613388
	0.312998306	10	9.389949	
	0.254414589	10	7.632438	
SENSITIVE AG	0.6582929	10	19.74879	20.133191
	0.6310982	10	18.93295	
	0.723928	10	21.71784	
SENSITIVE AP	0.23051636	10	6.915491	7.1780256
	0.2341	10	7.023	
	0.2531862	10	7.595586	
SENSITIVE DD	0.2364303	10	7.092909	5.5448901
	0.146831885	10	4.404957	
	0.171226823	10	5.136805	

Table 3. Hyaluronidase assay results

Sample	Skin Type	O.D of Sample	O.D of Control	% Inhibition
Dry Skin Formulations	Dry DD	0.0497	0.096	93.16%
	Dry AP	0.0659	0.096	45.68%
	Dry AG	0.0860	0.096	11.63%
Oily Skin Formulations	Oily DD	0.0565	0.096	69.91%
	Oily AP	0.0754	0.096	27.32%
	Oily AG	0.0560	0.096	71.43%
SensitiveSkin Formulations	Sens DD	0.0647	0.096	48.38%
	Sens AP	0.0654	0.096	46.79%
	Sens AG	0.0769	0.096	24.84%

Future research should focus on optimizing formulations to achieve consistent performance across all skin types.

4. DISCUSSION

The findings from this study highlight the potential of various formulations in addressing specific skin concerns, with particular focus on enzyme inhibition and sun protection factors (SPF). These results provide a foundation for tailoring skincare formulations to meet the unique needs of different skin types.

The results from the enzyme inhibition assays emphasize the variability in the efficacy of different formulations in inhibiting hyaluronidase and tyrosinase, enzymes associated with skin aging and hyperpigmentation. For dry skin formulations, the Dry DD formulation exhibited strong hyaluronidase inhibition, underlining its potential to maintain skin elasticity and hydration—two critical factors for addressing dryness. In contrast, the Dry AG formulation demonstrated the highest tyrosinase inhibition, suggesting its efficacy in reducing hyperpigmentation and promoting a brighter complexion.

For oily skin formulations, Oily AG and Oily DD formulations showed superior hyaluronidase inhibition, indicating their suitability for moisture retention and preventing premature aging in oily skin types. Oily AP exhibited the highest tyrosinase inhibition, making it a promising option for reducing pigmentation issues commonly associated with oily skin.

For sensitive skin formulations, Sens DD and Sens AP formulations displayed moderate hyaluronidase inhibition, while Sens AP demonstrated the most effective tyrosinase inhibition. These results suggest that these formulations could deliver gentle anti-aging and skin-brightening benefits while catering to the sensitivity of this skin type, ensuring efficacy without irritation. These findings underscore the potential for customizing formulations based on skin type to maximize efficacy. Future studies should explore the optimization of active ingredients in these formulations and their long-term stability and efficacy in clinical settings. The SPF values of the tested formulations further illustrate their potential for sun protection, which is critical for preventing UV-induced skin damage and supporting overall skin health. For dry skin formulations, the SPF analysis revealed that the Dry DD formulation had the highest SPF value (13.68), providing superior UVB protection among the dry skin-specific products. This highlights its potential for protecting dry skin against erythemal radiation. The SPF values for Dry AG and Dry AP formulations were 4.68 and 7.13, respectively, aligning with moderate protection levels suitable for everyday use. For oily skin, the AG formulation displayed the highest SPF (10.87), indicating its effectiveness in photoprotection for individuals with this skin type. The Oily DD and Oily AP formulations exhibited SPF values of 8.96 and 4.66, respectively, reflecting moderate to low UVB protection.

Sensitive skin formulations demonstrated the highest SPF values overall, with the AG formulation achieving an SPF of 20.13. This underscore sits strongUVB protection, meeting the higher photoprotection needs of sensitive skin types. The SPF values for Sens AP andSens DD formulations were 7.18 and 5.54, respectively. The SPF results align with existing literature on sunscreen products, which typically offer SPF values between 15 and 50. While the AG formulation for sensitive skin approaches industry benchmarks with its SPF of 20.13, there remains room for improvement to achieve equivalency with premium sunscreens offering broad-spectrum protection.

Comparing these formulations with commercial sunscreen products highlights their competitive efficacy while pointing to areas for enhancement. For example, industry-standard formulations containing zinc oxide and titanium dioxide typically achieve SPF values around 30–50, offering robust UVB protection alongside broad-spectrum coverage. The Detan & Detox formulation for dry skin, with an SPF of 13.68, aligns with lighter products marketed for daily use, such as moisturizers with added SPF in the range of 10–20. Similarly, the AG formulation for sensitive skin shows promise for optimization to meet higher benchmarks, combining photoprotectionwith minimal irritation.

Future work should focus on optimizing active ingredients in each formulation to enhance both enzyme inhibition and SPF values, benchmarking these formulations against industry-leading products to identify areas for further improvement, conducting in vivo studies to validate laboratory findings under real-world conditions, and exploring the development of formulations with broad-spectrum UV protection to address both UVA and UVB radiation. These steps will help bridge the gap between experimental formulations and commercially viable skincare solutions, ensuring efficacy, stability, and safety for various skin types.

ACKNOWLEDGEMENTS

We would like to extend our sincere thanks to all who have helped with the successful completion of this project. Firstly, we would like to sincerely thank Dr. Nalinkanth V Ghone, Associate Professor, Dept. of Chemical Engineering for providing us with the opportunity to do the project and for giving valuable input, constant support and educating us throughout the journey of this project. We would like to express our deep sense of gratitude to the Research Center of SSN College of Engineering, for providing us with the necessary funds for the completion of this project. We would like to express our sincere gratitude to Dr. Sathish Kumar, Head of the Department, Department of Chemical Engineering, for providing the facilities to do this project in our Department.

4. REFERENCES

1. Veerabadran, N. G., Price, R. R., & Lvov, Y. M. (2007). Clay nanotubes for encapsulation and sustained release of drugs. *Nano*, *2*(02), 115-120..
2. Nalinkanth V.Ghone, Dmitriy Mongayt,Vladimir Torchilin, Ronald R .Price,Yuri M. Lvov, "Organized Shells on Clay Nanotubes for Controlled Release of Macromolecules", Macromolecular Rapid

Communications, 2009, 30(2), 99-103.
3. Rawtani, D., & Agrawal, Y. K. (2012). Multifarious applications of halloysite nanotubes: a review. *Rev. Adv. Mater. Sci, 30*(3), 282-295.
4. Lohita S, Sriram Gokulakrishnan R, "Study of the controlled release of antioxidants from clay nanotubes", Mentor-Nalinkanth V. Ghone, UG Thesis, Anna University, April 2023.
5. Amit Roy, A. R., Sahu, R. K., Munglu Matlam, M. M., Deshmukh, V. K., Jaya Dwivedi, J. D., & Jha, A. K. (2013). In vitro techniques to assess the proficiency of skin care cosmetic formulations..
6. Rawtani, D., Pandey, G., Tharmavaram, M., Pathak, P., Akkireddy, S., & Agrawal, Y. K. (2017). Development of a novel 'nanocarrier'system based on Halloysite Nanotubes to overcome the complexation of ciprofloxacin with iron: An in vitro approach. *Applied Clay Science*, *150*, 293-302.
7. Kumar, L., Deshmukh, R. K., Hakim, L., & Gaikwad, K. K. (2024). Halloysite nanotube as a functional material for active food packaging application: a review. *Food and Bioprocess Technology*, *17*(1), 33-46.
8. Thalakkal, A. K., Muthukutti, G. P., Vellingiri, S., Naveen, S., & Madhu, S. (2023). Utilization of silicon enriched natural halloysite nano tube: mechanical, tribological, and corrosion behaviour in magnesium matrix composites. *Silicon*, *15*(11), 4823-4834.
9. Suh, Y. J., Kil, D. S., Chung, K. S., Abdullayev, E., Lvov, Y. M., & Mongayt, D. (2011). Natural nanocontainer for the controlled delivery of glycerol as a moisturizing agent. *Journal of nanoscience and nanotechnology*, *11*(1), 661-665.
10. Suh, Y. J., & Cho, K. (2015). Immobilization of nanoscale sunscreening agents onto natural halloysite micropowder. *Materials transactions*, *56*(6), 899-904.
11. Hári, J., Gyürki, Á., Sárközi, M., Földes, E., & Pukánszky, B. (2016). Competitive interactions and controlled release of a natural antioxidant from halloysite nanotubes. *Journal of colloid and interface science*, *462*, 123-129.
12. Baschieri, A., Amorati, R., Benelli, T., Mazzocchetti, L., D'Angelo, E., & Valgimigli, L. (2019). Enhanced antioxidant activity under biomimetic settings of ascorbic acid included in halloysite nanotubes. *Antioxidants*, *8*(2), 30.
13. Shao, L., Wang, X., Yang, B., Wang, Q., Tian, Q., Ji, Z., & Zhang, J. (2017). A highly sensitive ascorbic acid sensor based on hierarchical polyaniline coated halloysite nanotubes prepared by electrophoretic deposition. *Electrochimica Acta*, *255*, 286-297.
14. Abdullayev, E., & Lvov, Y. (2016). Halloysite for controllable loading and release. In *Developments in clay science* (Vol. 7, pp. 554-605). Elsevier.
15. Borrego-Sánchez, A., Sainz-Díaz, C. I., Perioli, L., & Viseras, C. (2021). Theoretical study of retinol, niacinamide and glycolic acid with halloysite clay mineral as active ingredients for topical skin care formulations. *Molecules*, *26*(15), 4392.
16. Lvov, Y. M., DeVilliers, M. M., & Fakhrullin, R. F. (2016). The application of halloysite tubule nanoclay in drug delivery. *Expert opinion on drug delivery*, *13*(7), 977-986.
17. Stavitskaya, A., Batasheva, S., Vinokurov, V., Fakhrullina, G., Sangarov, V., Lvov, Y., & Fakhrullin, R. (2019). Antimicrobial applications of clay nanotube-based composites. *Nanomaterials*, *9*(5), 708.
18. Sarruf, F. D., Contreras, V. J. P., Martinez, R. M., Velasco, M. V. R., & Baby, A. R. (2024). The scenario of clays and clay minerals use in cosmetics/dermocosmetics. *Cosmetics*, *11*(1), 7.

Research Paper

STRUCTURAL AND FUNCTIONAL CHARACTERIZATION OF ORF 2 PROTEIN FROM HEPATITIS E VIRUS

Sarathy.M[1], Miruthula.S[1], Selvaganapathy.V[1], Puniethaa Prabhu[2]

[1] *Undergraduate Student, Department of Biotechnology, K. S. Rangasamy College of Technology, Tiruchengode – 637215.*
[2] *Professor, Department of Biotechnology, K. S. Rangasamy College of Technology, Tiruchengode – 637215.*
**Corresponding Author: Sarathy M*
Email id: msarathi250@gmail.com

ABSTRACT

The ORF2 protein of Hepatitis E Virus (HEV) is an important part of the virus's morphogenesis and function as it assists in the assembly of the virus as well as the immune evasion. One such protein, which is 660 amino acids long, is reported in various forms, most importantly as an infectious form (ORF2i) as well as a glycosylated form (ORF2g) which are crucial in the understanding of HEV pathogenesis. Nonetheless, the description of the mechanisms through which the protein ORF2 functions in relation to the host and the life cycle of the virus is still perplexed. This work intends to explore the structural properties of the ORF2 protein: characterizing the protein in silico by predicting its structural modifications and functional roles during HEV infection. In particular, we will characterize the primary structure of the protein and relate that to its intended function, build a three dimensional model of the protein, locate its intrinsically disordered regions, and determine regions of the protein that are likely to be phosphorylated and which may be able to form regulatory functions. The results generated from this study should attempt to explain the biological activity of ORF2 protein in its regulation of functions of HEV and assist in discovering new ways to treat the hepatitis E virus.

KEYWORDS: *Hepatitis E Virus (HEV), ORF2 protein, Viral assembly, Immune evasion, Infectious form (ORF2i), Glycosylated form (ORF2g), Pathogenesis, Structural characterization.*

1. INTRODUCTION

The hepatitis E virus (HEV) is a serious problem in areas with poor sanitation, being the main cause for acute hepatitis around the globe. The hepatitis E virus is known to be an enveloped single-stranded RNA virus characterized by three open reading frames (ORFs) namely: ORF1, ORF2 and ORF3. Of these, the ORF2 protein is the most important, since it encodes the viral protein of the capsid which has a central role in the formation of infectious viral particles and immune avoidance strategies. The ORF2 protein contains 660 amino acids and is expressed in multiple forms such as infectious variant – ORF2i and glycosylated forms-ORF2g, which vary in their role in the pathogenesis of HEV. As it stands, the ORF2 protein is a subject of active research, particularly its structural function and its contribution to the viral life cycle through host interference. Emerging evidence suggests that not only does the ORF2 protein consist the structure of HEV but it has also been seen to be targeted by the immune system of the host as a major antigen.

However, the function of ORF2 protein is still unknown and recent studies have made important contributions to the knowledge regarding functions of ORF2 protein. For example, a previously discovered Arginine-Rich Motif (ARM) within the N-terminal region of ORF2 controls its lifecycle and interactions with host cells. This motif is essential for the nuclear translocation of ORF2, thereby affecting its function in modulating host antiviral responses, as well as the dual topology of ORF2, that enables the synthesis of both infectious and non-infectious forms of this protein, Such findings highlight the intricate nature of the functional roles played by ORF2 and its position as a key regulator of HEV biology. It has been shown that glycosylation at specific sites within the ORF2 protein influences its stability and immunogenicity, but the biological significance of these modifications in respect to HEV infection still needs to be clarified.

This research seeks to investigate the structural aspects of ORF2 protein using a wide range of bioinformatics techniques. A number of tools will be used to determine its linear sequence, construct its 3D forms, and find the regions that are expected to be partially flexible and able to undergo changes, as well as determine sites for expected post translational alterations such as phosphohrylation and others that are likely to be functionally important determinants. In particular, Expasy ProtParam will be employed to analyze molecular weight, isoelectric point and other physicochemical properties. We have estimated that the three-dimensional structure of the protein will be modelled using Modweb for homology modelling in order to have a pictorial view on 3 dimensional configuration and locate the important functional sites. Furthermore, IUPred 2A will be used for the prediction of regions of intrinsic disorder that may affect protein interactions, whereas, NetPhos 3.1 will predict sites for phosphorylation that are important in cascades of signal transduction. The findings generated from this study will help in understanding the biological function of the ORF2 protein including its role in the HEV infective cycle. In the end, this investigation intends to provide information that will ease the process of carrying out experiments aimed at trying to elucidate the therapeutic aspects of the targeted hepatitis E virus infection. At the beginning of this introduction, you describe your project, explain the importance of the ORF2 protein, the current limitations in characterization of its function, and the strategy you intend to adopt to study the protein using bioinformatic tools.

2. MATERIALS AND METHODS

A systematic bioinformatics analysis of the orf2 protein, beginning with amino acid evaluation to assess its composition and physicochemical properties. This is followed by 3d structure prediction to determine its spatial conformation and disorder region prediction to identify intrinsically disordered area s that contribute to protein flexibility and interactions. Next, phosphorylation prediction is conducted to locate potential regulatory phosphorylation sites, and finally, structure-based function prediction is performed to annotate the protein's functional roles based on structural and domain information.

2.1. Materials used

- Uniprot: The Unified Protein sequence and Function Database.
- Protparam: Compute the Physico -chemical Parameters for a Protein Sequence.
- Mod Base: Repository of annoted 3D Protein Models.
- IUpred: Protein Disordered prediction.
- Net phos: Prediction of phosphorylation sites.
- InterPro: Protein sequence classification and functional analysis.

Uniprot

Uniprot is a central repository for protein information, providing a wealth of data and tools to explore and analyze proteins. Access uniprot either through the website or programmatically. The knowledgebase is divided into swiss-prot, which is manually curated, and trembl, which is computationally analysed. Uniprot offers various tools for sequence search, protein classification, functional analysis, and structure prediction.

Protparam

Expasy is an extensive bioinformatics resource portal offering various tools for analysing protein sequences and structures. Access the Website to look through using the search bar or to browse by topic or tool for finding the specific tool to use. Enter your protein sequence in Fasta format and then select any additional parameters if required. Click submit on your analysis, and then see your results displayed on screen or as a file downloaded. Expasy provides tools such as protparam for protein parameter calculation, expasy-server for various sequence analysis tools, and databases like swiss-prot and trembl for protein sequences. Always read tool documentation, experiment with parameters, and use the help resources provided by expasy result & discussion.

Modweb

The Modweb procedure begins with the user submitting a protein sequence, which is then compared to known structures in a database to identify suitable templates. Using these templates, Modweb builds 3D models of the protein through comparative modelling techniques. The generated models are evaluated for quality, and users can access them for further analysis or visualization. This automated process integrates seamlessly with resources like ModBase to streamline protein structure prediction and research applications.

IUPred2A

IUPred2A is a web-based tool for predicting intrinsically disordered proteins (IDPs) and their regions, as well as conditionally disordered regions based on factors such as redox state and protein binding interactions. To use IUPred2A, researchers can input protein sequences by entering SWISS-PROT/TrEMBL identifiers, pasting amino acid sequences, or uploading FASTA files. The tool offers options for long and short disorder predictions, along with context-dependent analyses. Upon submission, IUPred2A generates scores for each residue, indicating the likelihood of disorder, with values above 0.5 suggesting disordered regions. The output includes graphical representations that aid in the interpretation of results, making IUPred2A a valuable resource for studying the functional implications of protein disorder in biological systems.

NetPhos 3.1

To use NetPhos, access the Website and enter your protein sequence in FASTA format. Enter your sequence and submit it. The tool will then scan through your sequence to predict sites of potential phosphorylation at serine, threonine, and tyrosine residues. Results are given as predicted sites with scores associated with them, indicating a degree of likelihood for phosphorylation.

InterPro

To use InterPro, access the website and input your protein sequence in FASTA format. Submit the sequence to InterProScan, which will dissect it against colorful protein hand databases. The results will display matches to protein families, disciplines, and functional spots, furnishing perceptivity into the protein's structure, function, and evolutionary connections.

3. RESULTS AND DISCUSSION

3.1. Results

3.1.1. Amino Acid Evaluation Using Protparam

Hepatitis E virus is a zoonotic pathogen of the Hepeviridae family that mainly causes acute hepatitis in countries with poor sanitation. It is estimated that one billion people worldwide are exposed to HEV, with millions infected. In India, HEV cases are often underreported since most patients seek traditional healing methods since treatment is rarely available in allopathic medicine. HEV is an important cause of acute hepatitis and liver failure with incidence rates of 10-40% and 15-45%, respectively. The virus has eight genotypes, and the primary human infections occur due to contaminated water by genotypes I and II. Genotypes III and IV are known to infect humans and animals alike and are known to cause chronic hepatitis in immunocompromised hosts. Cases in developed countries are often associated with raw meat consumption and contact with infected animals, which complicates the impact of HEV on public health. The viral genome of HEV consists of three defined open reading frames (ORFs): ORF1, ORF2, and ORF3. ORF1 encodes non-structural proteins that are necessary for viral replication. ORF2 encodes the capsid protein, which forms the major structural component of the virion.

The ORF2 protein is important for the integrity and assembly of viral particles upon infection and has both ordered and disordered regions. In ORF2, IDRs are those regions that do not form a stable 3D structure, thereby contributing to the evasion of the immune response. IDRs in viral proteomes, such as those of ORF2, play roles in disease processes and may represent potential drug targets. Thus ORF2, being highly involved in the binding of host cell molecules, their regulation, and viral assembly plus highly immunogenic, is being considered as a potential candidate as a vaccine against HEV. This study focuses its approach on the computational analysis of HEV's ORF2 protein disordered regions concerning its functional attributes, as it is involved in its pathogenesis, host, as immune evasion activity. Through explanation of molecular properties of ORF2 and, specifically its IDRs, this research study aspires to detect its heterogeneity in the HEV life cycles and provide an option for identifying therapeutic targets or potential vaccine candidates against HEV. Public health problems through HEV diversity: Because of this, different genotypes of the virus can have quite diverse transmissions routes with even zoonotic capabilities. This research will answer the need for the characterization of the structural and functional properties of ORF2, particularly its IDRs, which could explain the mechanisms through which HEV avoids host immunity and creates infection. The potential for ORF2 disordered regions as therapeutic targets should be addressed by this study and contribute to treatments or preventive measures against HEV infection.

Amino Acid Prediction of GT-I

From Table 1, The ORF2 protein from GT-I is composed of 660 amino acids with the molecular weight of 11,555.69 Da and theoretical pI of 9.00 suggesting that the protein is basic. High composition of amino acids, especially Pro (18.6%), Leu (11.5%), and Ser (10.6%) point toward its structural profile. It has three negative residues, Asp and 8 positive residues, Arg.There are 1,632 atoms in this molecule, which has a formula of C512H826N146O138S10. At 280 nm, the extinction coefficient is 500 M-1cm-1 when cystine pairs are formed and 0 when all the Cys residues are in the reduced state-the absorbance properties undergo drastic changes based on the cysteine state. Its estimated half-life is approximately 30 hours in mammalian reticulocytes, >20 hours in yeast, and >10 hours in *E. coli*, indicating moderate stability in vivo. Because the value for an instability index is 78.11, the protein is unstable. The aliphatic index is as high as 90.62, which could mean potential thermostability, but it has a GRAVY score of 0.391, indicating being hydrophobic.

Amino Acid Prediction of GT-II

From Table 1, The ORF2 protein of GT-II has 660 amino acids, with a molecular weight of 12,714.11 Da and theoretical pI 8.78; hence it is a basic protein by nature. Amino acid composition is dominated by Pro (17.9%), Leu (13.0%), and Ala (11.4%), which reflects a rather high proportion of residues with hydrophobic and apolar properties. It has 3 negatively charged residues (Asp + Glu) and 7 positively charged residues (Arg + Lys), reflecting charge balance. The atomic composition has 1,797 atoms and the molecular formula is C570H909N157O150S11.

Extinction at 280 nm is 6,000 M-1cm-1 if cystine pairs are formed, 5,500 M-1cm-1 when the Cys residues are reduced, thus depending on the cysteine status. The half life is estimated to be about 30 hours in mammalian reticulocytes, >20 hours in yeast and >10 h in

E. coli, hence moderately stable. However, 79.24 instability index classified the protein as unstable. It had an aliphatic index of 93.66 with satisfactory thermostability, while GRAVY score revealed 0.419 shows hydrophilic character.

Amino Acid Prediction of GT-III

From Table 1, the protein had 660 amino acids, a molecular weight of 12,375.64 Da, and a theoretical isoelectric point (pI) of 8.47 (suggested to be basic in nature). The absolute frequency of the richest amino acids was proline 18.0%, leucine 11.5%, and serine 12.3%, and all lysine, tryptophan, and tyrosine were absent. It has 3 negatively charged residues (Asp + Glu) and 6 positively charged residues (Arg), with a slight positive charge. The atomic composition of the protein totals at 1,735 atoms, with a molecular formula C546H875N149O152S13. Because of the absence of tryptophan, the extinction coefficient is low at 500 M^{-1} cm^{-1} with containing cystine formation and 0 M^{-1} cm^{-1} if all cysteines are reduced, causing an absorbance of 0.040 and 0.000, respectively, at 0.1% concentration. The calculated half-life is 30 hours in mammalian reticulocytes, over 20 hours in yeast, and over 10 hours in *E. coli*. A very high instability index of 90.57 classifies the protein as unstable, but an aliphatic index of 86.39 reveals good thermal stability. The positive GRAVY score of 0.473 further suggests a hydrophobic nature, which might influence its function and interactions inside the cellular environment.

Amino Acid Prediction of GT-IV

From Table 1, the ORF2 protein of HEV Genotype IV consists of 660 amino acids with a molecular weight of 11,671.82 Da and a theoretical pI of 8.60. The protein has the remarkable composition of proline (18.4%), leucine (13.2%), and alanine (10.5%) while showing no lysine or tryptophan residues and very low levels of negatively charged residues (3) and positively charged residues (6). The atomic formula is C524H838N142O141S9, accounting for 1654 atoms. The extinction coefficient at 280 nm of the thiol is 1865 M^{-1} cm^{-1} for cystine formation and 1490 M^{-1} cm^{-1} if all cysteines are reduced, with absorbances of 0.160 and 0.128, respectively. The estimated half-life is 30 hours in mammalian reticulocytes, over 20 hours in yeast, and over 10 hours in *E. coli.* Instability index is 78.73 with a classification that puts the protein as unstable, while the high aliphatic index of 95.00 suggest a significant thermostability. GRAVY score is 0.449; hence, it suggests a hydrophobic nature of the protein, which signifies lower solubility in an aqueous environment.

Amino Acid Prediction of GT-V

From Table 1, the ORF2 protein of HEV Genotype V has 660 amino acids. Its molecular weight is 11,353.54 Da, and the theoretical pI is 8.78. The most abundant amino acids are proline at 20.5%, leucine at 13.4%, and alanine at 9.8%. Amino acids absent are asparagine, tryptophan, lysine, and tyrosine. The protein contains a relatively high proportion of cysteine amounting to 7.1% and no ambiguous residues. The series has two negatively charged residues of Asp + Glu and six positively charged residues of Arg. Atomic composition is composed of 509 carbon, 824 hydrogen, 140 nitrogen, 135 oxygen, and 9 sulfur atoms totalling 1,617 atoms. Because there are no tryptophan residues, extinction coefficient is very low, at 500 M^{-1} cm^{-1} for cystine formation and zero for reduced cysteine with minimal absorbance at 280 nm. The estimated half-life is 30 hours in mammalian reticulocytes, over 20 hours in yeast, and over 10 hours in E. coli, thus reasonable stability in different systems; it is unstable, having an instability index of 84.80. The aliphatic index is 98.39, that means thermostability is high, and the GRAVY score of 0.480 shows hydrophobic nature and low solubility in aqueous environments.

Amino Acid Prediction of GT-VI

From Table 1, The ORF2 protein from GT-VI has 660 amino acids, weighing approximately 11,486.54 Da molecular weight, and theoretically has a pI of 8.53. It is dominated by Proline at 20.5%, Leucine at 11.6%, Glycine at 8.9%, and Valine at 8.9%, with no Asparagine, Lysine, or Tryptophan. It has 3 negatively charged residues (Asp + Glu) and 6 positively charged residues (Arg). The atomic composition of this protein yields a molecular formula of C511H811N143O138S10, which totals 1,613 atoms. The extinction coefficient is 1,990 M-1 cm-1 assuming cystine formation and 1,490 M-1 cm-1 if the cysteines are reduced. It has an estimated half-life of 30 hours in mammalian reticulocytes (in vitro), over 20 hours in yeast, and over 10 hours in *E. coli* (in vivo). An instability index of 81.50 places it under the unstable category; however, its aliphatic index of 82.68 hints towards moderate thermostability. A score for GRAVY is 0.238, which indicates a slight degree of hydrophobicity.

Amino Acid Prediction of GT-VII

From Table 1, The ORF2 protein of GT-VII contains 660amino acids, with a molecular weight of 11,518.58 Da and a theoretical isoelectric point (pI) of 8.78, showing minor positive charge at physiological pH. The amino acid composition has immense percentages of

Pro (15.9%), Leu (12.4%), and Ala (10.6%). There are only 2 negatively charged residues, (Asp + Glu) and 6 positively charged residues, (Arg + Lys). This indicates an overall modestly positive character. The atomic composition formula is C510H815N147O137S10, which equals 1,619 atoms. The extinction coefficient is cited as about 500 M^-1cm^-1, but be careful using this estimate since there are no tryptophan residues that could result in over 10% error in this estimation. An estimated half-life of 30 hours is seen in mammalian reticulocytes, >20 hours in yeast, and >10 hours in *E. coli*, which roughly amounts to being stable enough in all the systems, but this protein remains unstable as its instability index is 64.84. The aliphatic index is 92.39, thus suggesting good thermal stability and the GRAVY score is 0.438, which means it is hydrophilic; this could affect the solubility of the protein and interactions in biological systems.

Table 1. Amino Acid Distribution Pattern Prediction in Hev-Orf3 Sequences

AA	GT-1%	GT-2%	GT-3%	GT-4%	GT-5%	GT-6%	GT-7%	GT-8%
Ala	11.4	10.5	10.9	11.2	11.1	10.9	10.7	9.6
Arg	7.2	7.2	7.1	6.9	7.0	6.9	7.1	7.0
Asn	2.7	2.5	2.8	2.6	2.8	2.6	2.7	3.2
Asp	4.7	4.6	4.0	4.2	4.2	4.1	4.4	4.3
Cys	2.7	2.7	2.7	2.8	2.9	2.9	2.7	2.8
Gln	3.2	3.7	3.3	3.3	3.5	3.2	3.4	3.3
Glu	5.0	5.1	5.8	5.3	5.4	5.4	5.3	5.1
Gly	7.0	7.7	7.2	7.2	7.3	7.0	6.9	7.3
His	3.1	3.4	3.2	3.2	3.2	3.2	2.9	2.9
Ile	4.5	4.1	4.6	4.2	3.7	4.0	4.4	4.0
Leu	9.3	9.4	9.3	9.3	9.1	9.2	9.6	9.3
Lys	2.2	2.2	2.3	2.6	2.5	2.4	2.5	2.3
Met	1.3	1.2	1.2	1.2	1.2	1.3	1.4	1.2
Phe	4.4	4.4	4.2	4.2	4.6	4.6	4.3	4.3
Pro	7.0	6.8	7.6	7.1	6.6	7.3	7.2	7.1
Ser	6.3	6.4	5.9	6.6	6.5	6.3	6.7	7.2
Thr	6.4	6.4	6.0	6.0	5.9	6.6	6.0	6.0
Trp	1.5	1.5	1.6	1.6	1.5	1.6	1.6	1.6
Tyr	2.9	2.7	3.1	3.0	2.9	2.8	3.1	3.1
Val	7.5	7.4	7.2	7.6	8.0	7.6	7.3	8.2

Note: The amino acid values are mentioned as percentages.

From TABLE-1: GT I (JF443720); GT II (M74506); GT III (AB222182); GT IV (GU119961); GT V (AB573435); GT VI (AB602441); GT VII (KJ496143); GT VIII (KX387865).

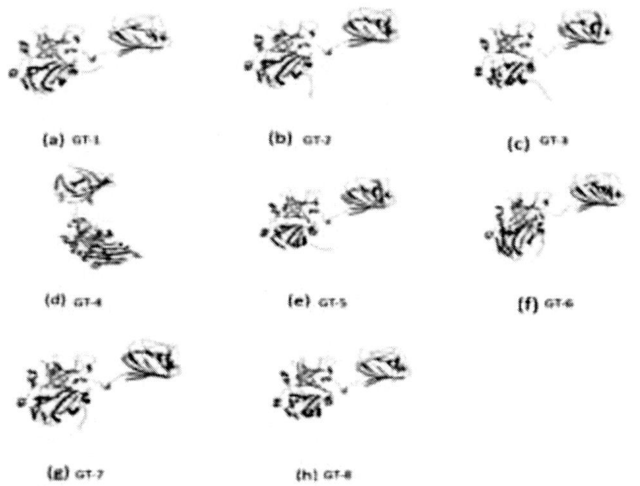

Figure 1.3D Structure of HEV-ORF2. (a) GT I (JF443720), (b) GT II (M74506), (c) GT III (AB222182), (d) GT IV (GU119961), I GT V (AB573435), (F) GT VI (AB602441), (G) GT VII (KJ496143) and (h) GT VIII (KX387865).

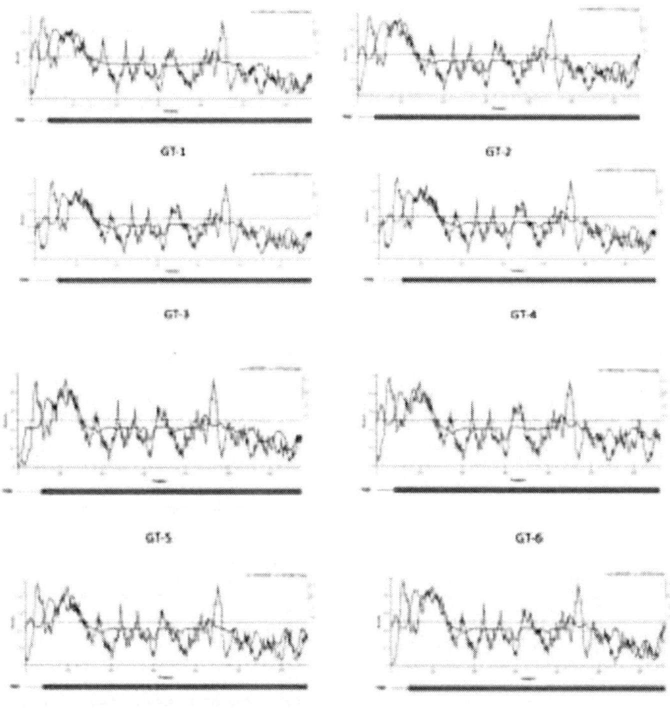

Figure 2.Intrinsic disorder analysis of HEV-ORF2. Intrinsic disorder distribution patterns depicted by graphs (A–H), (a) GT I (JF443720), (b) GT II (M74506), (c) GT III (AB222182), (d) GT IV (GU119961), I GT V (AB573435), (F) GT VI (AB602441), (G) GT VII (KJ496143) and (h) GT VIII (KX387865). The analysis was conducted through IUPred2A . Disorder probability was computed by setting 0.5 threshold values (dashed line). The regions above this threshold value are estimated as disordered.

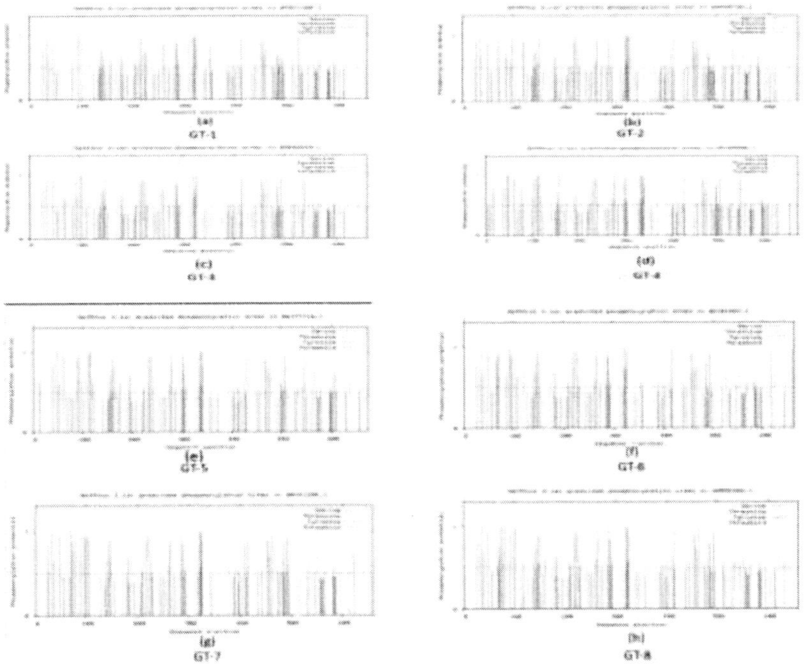

Figure 3. Identification of phosphorylation sites (Serine(Red), Threonine(Green), Tyrosine(Violet), Threshold(Threshold)) within HEV-ORF3, (A) GT 1 (JF443720), (B) GT 2 (M74506), (C) GT 3 (AB222182), (D) GT 4 (GU119961), (E) GT 5 (AB573435), (F) GT 6 (AB602441), (G) GT 7 (KJ496143) and (H) GT 8 (KX387865). The resulting score was computed using Netphos 3.1. The line (0.5 threshold value) is set to discriminate ordered and disordered residues

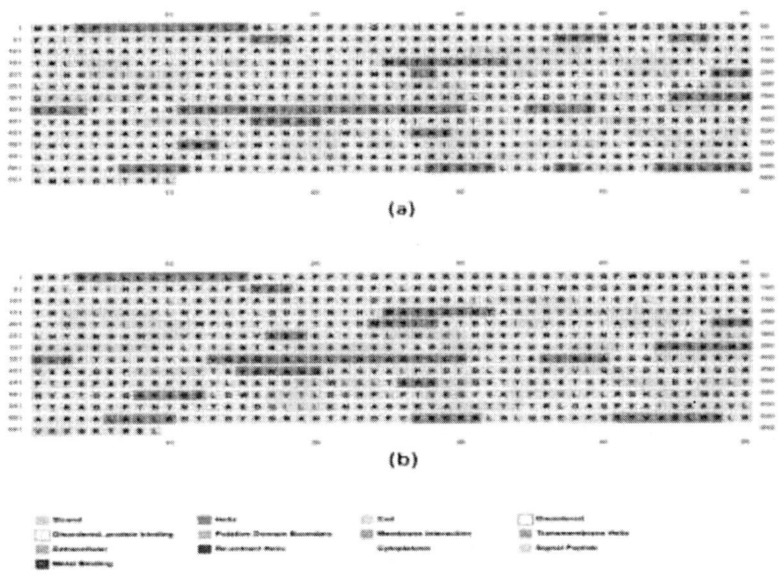

Figure 4. Structure Based Function by Identifying (Strand, Helix, Coil, Disordered, Putative Domain Boundary, Membrane Interaction, Transmembrane Helix, Extracellular, Re-entrant helix, Extracellular, Re-entrant Helix, Cytoplasmic, Signal Peptide, metal Binding) of ORF2 from HEV (a) GT 1 (JF443720), (b) GT 2 (M74506).

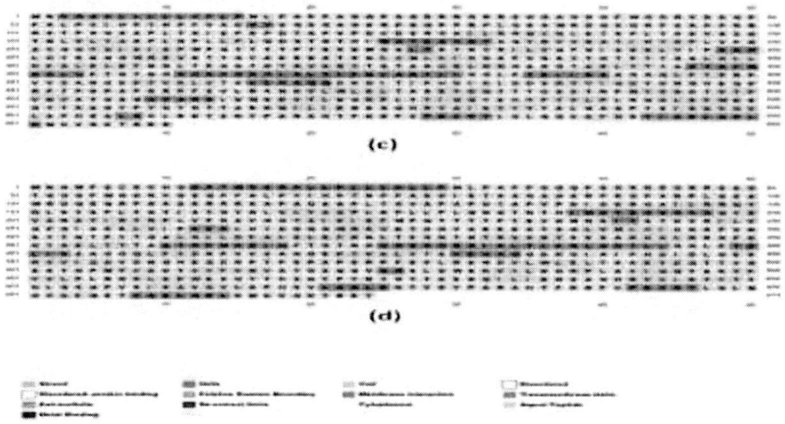

Figure 5. Structure Based Function by Identifying (Strand, Helix, Coil, Disordered, Putative Domain Boundary, Membrane Interaction, Transmembrane Helix, Extracellular, Re-entrant helix, Extracellular, Re-entrant Helix, Cytoplasmic, Signal Peptide, metal Binding) of ORF2 from HEV. (c) GT 3 (AB222182), (d) GT 4 (GU119961).

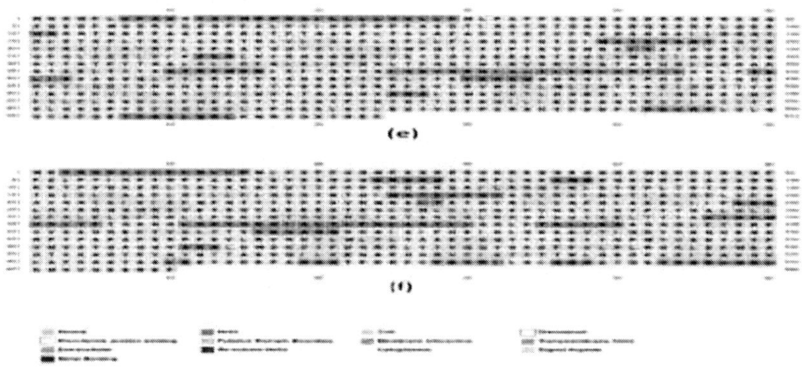

Figure 6. Structure Based Function by Identifying (Strand, Helix, Coil, Disordered, Putative Domain Boundary, Membrane Interaction, Transmembrane Helix, Extracellular, Re-entrant helix, Extracellular, Re-entrant Helix, Cytoplasmic, Signal Peptide, metal Binding) of ORF2 from HEV. (e) GT 5 (AB573435), (f) GT 6 (AB602441).

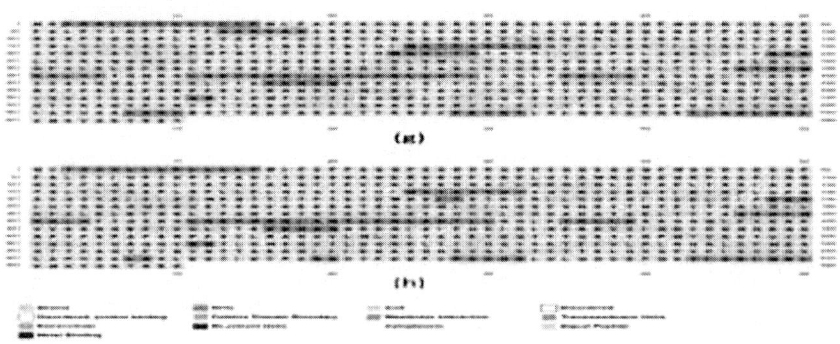

Figure 7. Structure Based Function by Identifying (Strand, Helix, Coil, Disordered, Putative Domain Boundary, Membrane Interaction, Transmembrane Helix, Extracellular, Re-entrant helix, Extracellular, Re-entrant Helix, Cytoplasmic, Signal Peptide, metal Binding) of ORF2 from HEV (g) GT 7 (KJ496143) and (h) GT 8 (KX387865).

Amino Acid Prediction of GT-VIII

From Table 1, The calculated molecular weight of the 660 amino acid ORF2 protein is 11,408.40 Da and has a theoretical isoelectric point or pI of 8.53. Thus, at physiological pH, it has a charge which is predominantly positive. Its amino acid composition shows that in addition to the 10.6 % of Ser residues, it contains more or less 14.2 % Leu and 15.0 % Pro residues. Of the total of 2 negatively charged residues (Asp + Glu), the number of positively charged residues (Arg + Lys) is 5, and the overall character is slightly acidic. Atomic formula C503H8139N1436O1415S9, with a total number of atoms of 1,609. The extinction coefficient is assumed to be 1,990 M-1cm-1, assuming all cysteine residues to be sulfide-bonded to each other (cystines), and 1,490 M-1cm-1, assuming all cysteines to be reduced. This corresponds to the protein's capability to absorb light at 280 nm. Estimated half-lives are approximately 30 hours in mammalian reticulocytes, >20 hours in yeast, and >10 hours in *E. coli*, the protein is reasonably stable in various biological environments, although it has an instability index of 58.91, and thus unstable. The aliphatic index of 99.29 means that the protein is thermostable, but the GRAVY score of 0.465 indicates a hydrophilic nature, which would cause probable insolubility and poor interactions within the biological environment.

3.1.2. 3D Protein Modeling Using Modweb

Modweb is known to be a strong technique for predicting the 3D structure of proteins directly from their amino acid sequences through 3D protein modelling. The Results are Shown in Figure 1, The comparative modelling process made by Modweb is semi-automated, during which the target protein sequence aligns with known protein structures, generally referred to as templates. These templates are then built into a 3D model. This is very useful when no experimental structures are available and offers insights into the function, interactions, and potential drug targets of the protein. Modweb combines many tools and databases, like MODELLER for building models and MODBASE for storage and retrieval of annotated protein models. The target protein sequence submission to Modweb initiates the process by searching suitable templates within the MODBASE database.

Once templates are identified, the server performs sequence-structure alignment and generates multiple models using different algorithms. These models are graded on quality and accuracy; the best-scoring models are chosen for further evaluation. The models can then be used by researchers in studying protein-ligand interactions, predicting the effects of mutations, and designing new drugs or therapeutic interventions. The ability to quickly and efficiently create accurate and reliable protein models makes Modweb a very useful resource for scientists to further the cause in structural biology, drug discovery, and biotechnology.

3.1.3. Disorder Prediction Using IUPred 2A

IUPred2A is such a sophisticated tool to predict the intrinsic disordered regions in the proteins, which is those areas that are missing their tertiary structure at the native environment. Such regions of high interest can be analyzed further in the protein sequence when it contains ordered as well as disordered regions such as HEV ORF2 proteins. The Results are enlisted in Figure 2, By analyzing the amino acid sequence of ORF2, IUPred2A can predict regions that do not form stable structures, which contribute to the virus's ability to evade the host immune response. The prediction of these disordered regions is important for understanding the functional attributes of ORF2, such as its involvement in pathogenesis, host interactions, and immune evasion. The IUPred2A tool is based on the biophysics model in providing a score for each residue of amino acid, which will imply the possibility of it forming part of a disordered region. Such information is highly crucial for characterizing structural and functional properties of ORF2, especially in terms of its IDRs, which are significantly involved in disease processes and thus represent targets for drug action. By explaining the molecular properties of ORF2, scientists can understand the versatility of ORF2 in the life cycle of HEV and discover therapeutic targets or vaccine candidates against HEV. The use of IUPred2A in this study underlines the importance of advancing our knowledge of viral proteins and their interactions with host cells.

- **Disorder Prediction of GT-1**

This graph shows intrinsic disorder and potential binding sites along the amino acid sequence of the ORF2 protein of HEV genotype I. The x-axis is amino acid positions and the y-axis is the disorder scores 0 to 1, where any value above 0.5 is considered a disordered region. The red line (IUPred2 long) is the prediction of disorder along the sequence, which shows high scores at the beginning and in certain peaks across the protein, suggesting a large number of flexible, unstructured regions in ORF2. The blue line (ANCHOR 2) is the probable binding sites within these disordered regions, with peaks suggesting areas where ORF2 may be involved in interactions, likely because of the structural flexibility of the protein. Such an interaction potential is in general typical of intrinsically disordered proteins that allow the nonspecific interaction of a single protein molecule with multiple partners in a transient manner. A bar at the bottom of the PFAM shows the conserved domain of ORF2 marked in pink, indicated to be evolutionarily conserved and functionally important in HEV. Overall, these data suggest that the ORF2 protein of GT-I contains large portions of disordered regions and potential binding sites, which could be responsible for plastic interactions and functional roles in the context of the viral lifecycle.

- **Disorder Prediction of GT-2**

This graph represents the intrinsic disorder prediction of ORF2 protein from HEV genotype II: x-axis – amino acid positions, y-axis is the disorder score. The disorder score of IUPred2 long is the predicted disorder along the sequence, with scores above 0.5 indicating disordered regions, particularly notable near the start and with peaks scattered throughout. It suggests that considerable parts of ORF2 are flexible and unstructured. The blue line AnCHOR2 suggests the presence of probable binding regions in such disordered segments, as peaks represent areas of molecular contact, which is a very common feature of disorderly proteins. The bottom PFAM bar marks a conserved domain (pink), indicating that it is evolutionary stable and functional in the context of the ORF2 sequence. Overall, this suggests that the ORF2 of GT-II has a very large area of disorder containing putative binding sites and, therefore, flexible functional and interaction capabilities probably contributing to its roles in the viral lifecycle.

- **Disorder Prediction of GT-3**

The graph is for intrinsic disorder prediction of the ORF2 protein from the HEV genotype III, and x-axis corresponds to the amino acid positions, whereas the y-axis is the disorder scores. The red line is IUPred2 long, showing regions of disorder, where a score >0.5 indicated unstructured flexible segments, highly dominating at very beginning and dispersed throughout the sequence. The blue line (ANCHOR 2) indicates possible binding sites within these disordered regions. These peaks represent probable interaction sites, a common characteristic of disordered proteins that form transient interactions between molecules. It points to the PFAM bar at the bottom with a conserved domain in pink, which suggests stability and functional importance in the course of ORF2 protein evolution. Overall, these data indicate large sections of the ORF2 of GT-III are unfolded structures with binding potential that yield conformational flexibility and interaction capabilities probably critical during the viral life cycle.

- **Disorder Prediction of GT-4**

The graph shows two lines representing "AR (1) Coef" and "LP Norm" over a sequence of positions. Both lines tend to decline with some fluctuations. The "AR (1) Coef" starts around 0.8 and declines to 0.2, which means that the correlation between consecutive values weakens. The "LP Norm," which was initially around 0.2, drops to a minimum around 0.1 at position 400, indicating a decrease in data variability. However, without more specific context, the interpretation is quite vague. Knowing the nature of data and the positions would have served to make the analysis more straightforward.

- **Disorder Prediction of GT-5**

This graph represents the protein sequence of ORF2 from GT5, with two different prediction methods applied: ANCHOR2 and IUPred2. The y-axis represents the score assigned by each method, indicating the likelihood of a region being disordered. The x-axis represents the position within the protein sequence. Both methods reveal a general trend of decreased disorder toward the end of the sequence. However, regions with higher disorder predicted by both methods are found, particularly at positions 200-300 and 500-600. Regions such as these could be important functionally and should be explored in greater detail.

- **Disorder Prediction of GT-6**

A graphical representation of the protein sequence of ORF2 from GT6 is given using two different prediction methods: ANCHOR2 and IUPred2. Each y-axis represents the score assigned by a respective method, which signifies the propensity for that region to be disordered, while the x-axis corresponds to the position in the protein sequence. Both methods show a general trend of decreasing disorder toward the end of the sequence. However, there are regions of higher disorder predicted by both methods, particularly around positions 200-300 and 500-600. These regions might be functionally important and warrant further investigation.

- **Disorder Prediction of GT-7**

The graph gives a visual representation of the protein sequence of ORF2 from GT7 with two different prediction methods applied: ANCHOR2 and IUPred2. The y-axis represents the score assigned by each method, indicating the likelihood of a region being disordered. The x-axis represents the position within the protein sequence. Both methods exhibit a general trend of decreasing disorder towards the end of the sequence. However, regions of higher disorder are predicted by both methods, especially around positions 200-300 and 500-600. These regions might be functionally important and should be investigated further.

➢ **Disorder Prediction of GT-8**

The graph illustrates the protein sequence of ORF2 from GT8 using two different prediction methods: ANCHOR2 and IUPred2. The y-axis represents the score assigned by each method, which shows the likelihood of a region being disordered. The x-axis represents the position within the protein sequence. Both methods exhibit a general trend of decreasing disorder at the end of the sequence. However, regions of higher disorder are predicted by both methods, notably at positions 200-300 and 500-600. These regions could be of functional importance and require further analysis.

3.1.4. Phosphorylation Prediction Using NetPhos 3.1

NetPhos 3.1 is a powerful bioinformatics tool designed to predict phosphorylation sites in eukaryotic proteins. The server uses ensembles of neural networks to predict the likelihood of serine, threonine, or tyrosine residues being phosphorylated. It supports both generic and kinase-specific predictions, making it versatile for a wide range of applications. The score for every residue in an amino acid sequence of a protein is given by NetPhos 3.1 based on its analysis. This is important information because phosphorylation is involved in major cell processes, such as signal transduction, regulation of protein-protein interactions, and enzyme activities.

The NetPhos 3.1 server is interactive; users can input a protein sequence in FASTA format, and for every residue, get a detailed prediction of potential phosphorylation. This information consists of the sequence context, scores of prediction, and which kinase is involved if the latter applies. The entire data thus helps to understand potential phosphorylation sites and their biological relevance. Providing accurate and reliable predictions, NetPhos 3.1 allows the study of post-translational modifications and their effects on the function of proteins, furthering knowledge in areas such as molecular biology, biochemistry, and drug discovery. The Results are shown in Figure 3.

➢ **Phosphorylation prediction of GT-1**

The number of sites predicted for phosphorylation by the phosphorylation prediction analysis of protein sequence AFB71105.1 was a total of 81 sites. Out of these sites, 37 sites, or about 46%, are the ones predicted to have a likelihood of being phosphorylated, whereas 44 sites, or about 54%, had no prediction and hence is not phosphorylated, as labelled with a dot in the "Answer" column. Conspicuous of the high-scoring phosphorylation sites is the one located at position 34 (context "RGRRSGGSG", Score: 0.996) and the one at position 190 (context

"VARATIRYR", Score: 0.912), where several kinases like PKC, GSK3, and INSR are implied in their phosphorylation. General distribution of phosphosites within the protein brought out some putative clusters that could suggest a regulatory function within the functional landscape. Some sites with scores greater than 0.4 were missed and they did not predict phosphorylation, meaning that something other than the score accounts for a larger contribution to the likelihood of phosphorylation. In conclusion, the present results give a great insight into how our comprehension of cellular processes through phosphorylation and regulatory mechanisms, especially the functional implications of these phosphorylation sites in contexts linked to disease, increases enormously.

➢ **Phosphorylation prediction of GT-2**

NetPhos 3.1b analysis of the ORF2, genotype 2 of HEV suggests a high degree of post-translational modification, as several phosphorylation sites can be seen along the sequence. Some serine and threonine, plus tyrosine residues are marked by GSK3, PKC, cdc2, etc. The high-confidence phosphorylation events, high scores, include S at 26, 34, 48, and 149; T at 37, 104, and 221; and Y at 182. What is noteworthy is that these sites indicate a regulatory function in protein-protein interactions and signal transduction. The unsp kinases, which occur quite frequently, suggest new or less-well-characterized kinase interactions. Clusters of phosphorylation sites, especially at the C-terminal and central regions, may influence protein conformation and thus its functionality, thereby influencing the engagement of the ORF2 protein in viral assembly, host interactions, and immune evasion mechanisms. These results point to functional sites with potential biochemical implications regarding the role of ORF2 in the viral lifecycle.

➢ **Phosphorylation prediction of GT-3**

NetPhos 3.1b was used to predict where ORF-2 of Hepatitis E Virus genotype III, 660 amino acids in length, is phosphorylated. Here, various serine (S) and threonine (T) residues along the sequence of protein appeared. Most of these sites would be expected to be phosphorylated by several kinases, some of which are less well characterized, including the unclassified (unsp) kinases, PKC, PKA, cdc2, GSK3, CaM-II, DNAPK, CKII, SRC, EGFR, PKG, and INSR. The scores for the prediction confidences were not the same and had relatively high scores for some sites (34, 48, 91, 102, 103, 216, 221, 226, 283, 537, 660). In these cases, the corresponding sites may potentially have high-level phosphorylation activities. The overall distribution of the phosphorylation sites around the protein is homogeneous and reveals regions of high serine and threonine density. The existence of several phosphorylation sites can imply a complex regulation by phosphorylation of the ORF-2 protein, which might affect functionalities, interactions, or the role of the viral life cycle. Regions between residues 50 to 700 are highly dense in phosphorylation sites, emphasizing the importance of post-translational modification to modulate the activity and stability of the protein.

➢ **Phosphorylation prediction of GT-4:**

The ORF-2 protein of the Hepatitis E Virus genotype IV sequence ACY66855.1 with 674 amino acids was screened for phosphorylation sites using the NetPhos 3.1b tool. In this sequence, several serine (S) and threonine (T) residues were found to be possible phosphorylation sites for phosphorylation throughout the entire sequence. Others that would phosphorylate such sites include unclassified (unsp) kinases, Protein Kinase C (PKC), Protein Kinase A (PKA), cyclin-dependent kinase 2 (cdc2), Glycogen Synthase Kinase 3 (GSK3), Calcium/Calmodulin-dependent protein kinase II (CaM-II), cyclin-dependent kinase 5 (cdk5), Casein Kinase II (CKII), Src kinase (SRC), Epidermal Growth Factor Receptor (EGFR), and Protein Kinase G (PKG). The confidence scores for the predictions for these sites varied, but some sites gave very high scores, such as positions 48, 49, 51, 105, 111, 112, 116, 335, 338, 339, 340, 503, 507, 531, and 551, for which the level of confidence for these phosphorylation events was strong. The phosphorylation sites appear to be spread out in the protein across certain regions showing densities for both serine and threonine residues. Multiple potential sites for phosphorylation indicate complex regulation that involves the phosphorylation of ORF-2 protein, which might influence the function and interactions or its contribution to viral life cycle events. Particular consideration should be given to the density of phosphorylation between residues 50 and 700, possibly an impact on stability and activation of the protein.

➢ **Phosphorylation prediction of GT-5**

For the protein of HEV (BAJ77116.1), ORF3, the results of NetPhos 3.1b prediction showed 12 sites of phosphorylation, with different types of amino acids-serine, threonine, and tyrosine-being involved as potential targets for modification. Serine residues at positions 48 with 0.996, position 90 with 0.911, and position 195 with 0.699 gave the best scores, which in most cases indicate a good degree of the potential for phosphorylation. Sites for the kinases include PKC, GSK3, and cdc2, among others, "YES" meaning verified phosphorylation potential. The protein has 674 amino acids with motifs corresponding to potential roles in the viral process. Overall, the data support the idea that phosphorylation plays an essential role in the ORF3 protein regulation and function, determining its multifunctionality in the process of HEV infection.

➢ **Phosphorylation prediction of GT-6**

Using the NetPhos-3.1b program to analyze the phosphorylation site of the ORF3 protein from HEV genotype IV showed multiple phosphorylation sites across 660 amino acids. These include residues S, T, and Y, which may be targeted for phosphorylation by many kinases, like PKC, GSK3, cdc2, and PKA to name a few. There were a number of significant phosphorylation sites, particularly on serine and threonine, scoring above the threshold for phosphorylation and showing associations with distinct kinases. Residues like S26, S34, S48, T103, T120, T216, and S657 scored very high values and thus presented tremendous potential for phosphorylation. Phosphorylation sites tended to cluster into specific regions that might be functionally relevant. They included unsp (unspecified) kinases all the way to those involved in cell signaling pathways, such as PKC and cdc2. The analysis would give the multifunctionality aspect of the ORF3 protein with regard to post translational modifications that might affect its function in the viral cycle and in contact with the host cells.

➢ **Phosphorylation prediction of GT-7**

Prediction results of the NetPhos-3.1b algorithm for the protein sequence AHY61296.1 with a length of 660 amino acids show various phosphorylation sites distributed along the sequence. Scanned, among others, are serine (S), threonine (T), and tyrosine (Y) residues. Many of these are determined to be potential sites of phosphorylation with scores greater than 0.5. For instance, there are numerous predicted phosphorations in the sequence with scores > 0.9: S25, S26, S34, which have strong indications of kinase interaction, especially with PKC and GSK3 kinases. Other identified residues like S55, S59, and S67 have poorer scores but do not have so much potential for the kinase interaction without any very strong indications of phosphorylation. Other residues as T140, T146, S145, and Y443 also have high scores thus increasing the complexity of the protein's phosphorylation. The extensive pattern of predicted phosphorylation sites therefore suggests potential regulatory roles in various cellular signaling pathways, making AHY61296.1 an important protein for further functional analysis relative to pathology caused by hepatitis E virus (HEV).

- **Phosphorylation prediction of GT-8**

Total number of the possible sites of phosphorylation was identified within 660 amino acids contained in ORF2 of GT-VIII by the use of netphos-3.1b: mostly S/T residues with a few Y sites. The number of confirmed sites was 61 YES while that of not confirmed sites were 35. The most abundant kinases corresponding to these sites were Protein Kinase C with 19 sites, Glycogen Synthase Kinase 3 with 14 sites, and cell division cycle 2 with 12 sites. Other kinases, such as Protein Kinase A, Calcium/Calmodulin-dependent protein kinase II, Epidermal Growth Factor Receptor, and Insulin Receptor Substrate, were also incriminated but at much lower frequencies. The prediction scores were found to range between 0.382 and 0.997, suggesting various degrees of confidence associated with the prediction of phosphorylation sites. Intriguingly, several high-confidence sites with scores >0.9 appear to correspond to unsp kinases, suggesting there could be regulatory importance to these sites. Overall, this thorough profiling pokes at the complexity of the ORF 2 of GT VIII post-translational modification and identifies major kinase interactions that should be significant for its functional regulation.

3.1.5 Structure-Based Function Prediction Using InterPro

InterPro is a rich resource for the functional analysis of protein sequences through classification into families and the prediction of domains and important sites. InterPro is the integration of protein signatures from multiple member databases into a single, searchable resource that leverages individual strengths to produce a powerful integrated database and diagnostic tool. In identifying protein families, domains, and critical sites, this uses predictive models referred to as signatures, and hence comes with detailed functional annotations relevant for Gene Ontology. Alltogether, this integration permits automated annotation of millions of sequences which help in the study of protein function and interaction.

InterPro utilizes a structure-based function prediction that requires one to submit a protein sequence for searching within its member databases for matching signatures. Following the identification of these signatures, InterPro generates a complete analysis of the protein, such as its family classification, domain architecture, and important functional sites. This information is extremely important to understand the function of this protein in biological processes and its interaction with other molecules or its potential as a target for drug development. It enables the integration of various sources' data into the platform through InterPro, allowing robust structural information-based predictions for identifying protein function in biological pathways.

The Results are shown in Figure 4,5,6 & 7 Computational analysis of the Hepatitis E contagion ORF2 protein with colourful bioinformatics tools gave great perceptivity into its structural and functional parcels. The amino acid evaluation with ProtParam revealed that there are ordered and disordered regions in the ORF2 protein, where naturally disordered regions were set up to play a major part in vulnerable elusion and pathogenesis. This finding stressed the significance of IDRs in the viral life cycle and their implicit as a remedial target. 3D protein modeling with Modweb successfully prognosticated the three- dimensional structure of the ORF2 protein, and detailed models are produced for farther studies. The quality and delicacy of the generated models were estimated, and the best- scoring models were named for analysis. These models are veritably precious for understanding the function of the protein, its relations, and implicit medicine targets. therefore, the mileage of Modweb in structural biology and medicine discovery is well demonstrated.

3.2 Discussion

This paper gives a comprehensive analysis of the hepatitis E virus (HEV) ORF2 protein, focusing on its structural and functional properties using various bioinformatics tools. The evaluation of amino acids using protparam showed that the ORF2 protein contains both ordered and disordered regions, with intrinsically disordered regions playing a crucial role in immune evasion and pathogenesis. These findings underscore the importance of IDRs in the viral life cycle and their potential as therapeutic targets.

This has allowed for the prediction of three-dimensional structures of the ORF2 protein with high quality models for further studies by applying 3D protein modelling on Modweb. The produced models were assessed for quality and accuracy, and the best-scoring models were taken into consideration for analysis. Such models provide valuable insight into the function, interaction, and potential drug target of the protein, therefore proving the utility of Modweb in structural biology and drug discovery. Iupred2a prediction of disorder in the protein ORF2 indicated particular regions that are not stable and therefore confer the virus with immune evasion capability. The biophysics based model provided scores for every amino acid residue in the protein, which showed its propensity to be part of a disordered region.

The Netphos 3.1 tool predicted phosphorylation sites in the ORF2 protein, which could explain its regulation and interactions. The neural network-based predictions of the tool indicated that serine, threonine, and tyrosine residues are more likely to be phosphorylated, which is crucial for many cellular processes. Thus, these results contribute to a better understanding of the post-translational modifications of ORF2 and their effects on HEV infection and replication. Overall, this study has brought out the importance of the ORF2 protein in the life cycle of HEV and its potential as a therapeutic target. Further experimentation, as well as exploration into the role of ORF2 in HEV pathogenesis and immune evasion, would further reinforce findings toward effective treatments and prevention of HEV infection.

LIST OF ABBREVIATIONS

1. **HEV**- Hepatitis E Virus
2. **AA** - Amino Acids
3. **UNIPROT** – Universal Protein Resource
4. **Modweb** – Comparative Protein Signature Database
5. **ORF** - Open Reading Frame
6. **PROTPARAM** – Protein Parameter Files.

REFERENCES

1. Zhao, M., Li, X. J., Tang, Z. M., Yang, F., Wang, S. L., Cai, W., ...& Zheng, Z. Z. (2015). A comprehensive study of neutralizing antigenic sites on the hepatitis E virus (HEV) capsid by constructing, clustering, and characterizing a tool box. *Journal of Biological Chemistry*, *290*(32), 19910-19922.
2. Chen, Y., Tang, Y., Zhang, S., Tian, Y., Xu, S., Zhang, C., ...& Liu, B. (2024). Evaluation of novel synthetic peptides of avian hepatitis E virus ORF2 as vaccine candidate in chickens. *Virus Research*, *349*, 199459.
3. Wang, B., & Meng, X. J. (2021). Structural and molecular biology of hepatitis E virus. *Computational and Structural Biotechnology Journal*, *19*, 1907-1916.
4. Liu, Z., Behloul, N., Baha, S., Wei, W., Tao, W., Zhang, T., ...& Meng, J. (2020). Role of the C-terminal cysteines in virus-like particle formation and oligomerization of the hepatitis E virus ORF2 truncated proteins. *Virology*, *544*, 1-11.
5. Ade, C. M., Derbes, R. S., Wagstaff, B. J., Linker, S. B., White, T. B., Deharo, D., ... & Roy-Engel, A. M. (2018). Evaluating different DNA binding domains to modulate L1 ORF2p-driven site-specific retrotransposition events in human cells. *Gene*, *642*, 188-198.
6. Gordeychuk, I., Kyuregyan, K., Kondrashova, A., Bayurova, E., Gulyaev, S., Gulyaeva, T., ...& Mikhailov, M. (2022). Immunization with recombinant ORF2 p551 protein protects common marmosets (Callithrix jacchus) against homologous and heterologous hepatitis E virus challenge. *Vaccine*, *40*(1), 89-99.
7. Shafat, Z., Ahmed, A., Parvez, M. K., & Parveen, S. (2023). Intrinsic disorder in the open reading frame 2 of hepatitis E virus: a protein with multiple functions beyond viral capsid. *Journal of Genetic Engineering and Biotechnology*, *21*(1), 33.

8. Eskandarzade, N., Ghorbani, A., Samarfard, S., Diaz, J., Guzzi, P. H., Fariborzi, N., ...& Izadpanah, K. (2022). Network for network concept offers new insights into host-SARS-CoV-2 protein interactions and potential novel targets for developing antiviral drugs. *Computers in biology and medicine*, *146*, 105575.
9. Tyagi, S., Korkaya, H., Zafrullah, M., Jameel, S., & Lal, S. K. (2002). The phosphorylated form of the ORF3 protein of hepatitis E virus interacts with its non-glycosylated form of the major capsid protein, ORF2. *Journal of Biological Chemistry*, *277*(25), 22759-22767.
10. Han, Q., Zhou, C., Wu, S., Liu, Y., Triplett, L., Miao, J., ...& Zhao, B. (2015). Crystal structure of Xanthomonas AvrRxo1-ORF1, a type III effector with a polynucleotide kinase domain, and its interactor AvrRxo1-ORF2. *Structure*, *23*(10), 1900-1909.
11. Xing, L., Li, T. C., Mayazaki, N., Simon, M. N., Wall, J. S., Moore, M., ...& Cheng, R. H. (2010). Structure of hepatitis E virion-sized particle reveals an RNA-dependent viral assembly pathway. *Journal of Biological Chemistry*, *285*(43), 33175-33183.
12. Kinast, V., Klöhn, M., Nocke, M. K., Todt, D., & Steinmann, E. (2022). Hepatitis E virus species barriers: seeking viral and host determinants. *Current Opinion in Virology*, *56*, 101274.

Research Paper

EXPLORING AQUEOUS PHASE OF WHEAT STRAW FROM PYROLYSIS AS AN EFFECTIVE BIOCONTROL AGENT AGAINST *RALSTONIA SOLANACEARUM– IN SILICO APPROACH*

Srividhya Krishnan [1,2], Saravanan Ramiah Shanmugam [2,3], Nithyanand Paramasivam[1*], Ponnusami Venkatachalam [2*]

[1]Biofilm Biology Laboratory, Centre for Research on Infectious Diseases (CRID), School of Chemical and Biotechnology, SASTRA Deemed University, Thanjavur, TN, 613401
[2]Biomass, Bioenergy and Bioproducts Laboratory, Centre for Bioenergy, School of Chemical and Biotechnology, SASTRA Deemed University, Thanjavur, TN, 613401
[3]Department of Biosystems Engineering, Auburn University, Auburn, AL, USA
***Corresponding Author**: Ponnusami Venkatachalam*
***Email id**: vponnu@chem.sastra.edu*

ABSTRACT

Ralstonia solanacearum is a wilt causing pathogen infecting tomato, potato and banana fields creating significant economic losses to farmers across the globe. They remain viable as they colonise through xylem tissues thereby decreasing the flow throughout the system. They are operated by the Phc-quorum sensing (QS) system, with *PhcA* and *PhcR* genes playing an important role in modulating the virulence in host environment. New compounds as biocontrol agents to tackle this pest is now gaining agricultural importance. The present work investigates on exploring the potential of the compounds present in the aqueous phase (AQ), obtained through pyrolysis of wheat straw (WS) against *Ralstonia solanacearum*.The compounds of the WS-AQ were screened using GC-MS analysis. The protein structures (PhcA and PhcR) were modelled, validated, prepared and docked against the compoundsidentified in aqueous phase via. GC-MS analysis.Molecular docking study confirmed that eight compounds showed drug-likeliness property. Among the variety of compounds tested both phenol and 2, 3-Dimethoxy phenol gave a docking score of -4.588 Kcal/mol and -3.934 Kcal/mol for PhcA and PhcR, respectively. Based on our previous study we can predict that an additive effect of these compounds in the AQ was responsible for the anti-pathogen activity. The results from this study are helpful towards developing a novel bio-based weedicide against this invasive pathogen.

KEYWORDS: *PhcA, PhcR, Pyrolysis, wilt disease, aqueous phase.*

1. INTRODUCTION

Invasion of disease-causing pathogenic microorganisms in plants such as tomato, potato, and banana is causing significant concern of-late. They infect the plants causing significant reduction in the crop yield, quality, and economic loss posing problems to the food security[1]. One such deadly pathogen is *Ralstonia solanacearum,* a gram-negative, rod-shaped soil born pathogen, which penetrates through the vascular system of the plants. It was estimated to have caused an annual loss of $950 million dollar, about 1.7 MH is affected by *R. solanacearum* in the potato fields[2]. The mode of infection is initiated via. sensing and reaching the roots of the plant by chemotaxis, followed by attachment via type IV pili. Then they enter through the opening and reaches the xylem within 24 hours by twitching motility thereby filling the entire vessels with biofilm matrix to restrict the flow of water[3]. The quorum sensing (QS) system of *R. solanacearum* is a well-established one. The Phc-QS system comprises of *PhcB, PhcS and PhcR* genes. The signals are sensed by histidine sensor (*PhcS*), transferred to transcriptional regulator (*PhcA*) which in turn activates the *PhcR*[4]. The currently developed control strategies include the physical (like bio-fumigation), chemical (pesticides), biological (using natural products) and crop rotation methods[5]. Out of all the methods, the urge for the development of eco-friendly and cost-effective approach for the defence system is in the look out.

Pyrolysis, is a thermochemical conversion process, that takes place in the absence of oxygen converts the biomass into three products viz., solid (biochar), liquid (Bio-oil aqueous phase BOAP) and gas (pyro-gas)[6]. The liquid by-product can be further separated into two phases, the oily and aqueous phase. The oil phase due to its high calorific value can be used in the fuel blends and other energy applications[7]. While, the aqueous phase is considered as waste and released into the water bodies[8]. Previously literatures have reported the use of the pyrolysis BOAP as insecticidal agents against different plant pests[9,10]. Wheat being India's second largest growing crop variety, post-harvest produces straw waste in major[11]. Studies from our group have reported the pyrolysis aqueous phase of wheat straw residue (WS AQ) were effective in disrupting the Methicillin resistant *Staphylococcus aureus* (MRSA) biofilms in the clinical settings[12]. Similarly, an *in silico* study on compounds present in the AQ obtained from *Parthenium hysterophorus* (an environmental weed) showed effective antifungal activity against fungal pathogen *Cryptococcus neoformans*[13].

Based on our previous findings and the existing literatures, we envisage that the aqueous phase (AQ) obtained from pyrolysis of wheat straw (WS) residue could be effective against the plant pathogen, *R. solanacearum.* Hence, the objectives of this study are as follows:

1. To perform protein structure modelling and validation of PhcA and PhcR of *R. solanacearum*
2. Toperform*In silico* docking and evaluate the docking scores for compounds present in the WS AQ
3. To perform molecular dynamics simulation for the selected protein-compound complex.

2. MATERIALS AND METHODS

2.1 Protein structure prediction and modelling of PhcA and PhcR

The 3D structure of both the proteins (PhcA and PhcR) are unavailable in the protein database (PDB)[14], hence the amino acid sequences are fetched from UniProt database and modelled using online servers like Alphafold and Itasser[15].

2.2 Structure validation

The modelled structures are validated using Ramachandran plot. The structures obtained from Alpha fold and Itasser were submitted in online server known as SAVES (Procheck) and the Ramachandran plots were generated. The best model was t considered for future analysis.

2.3 Protein structure preparation

The validated 3D structures (.pdb files) were subjected to protein preparation wizard (PPW), where the structures arepre-processedto identify any error with the structure like the missing hydrogen atoms, missing chains/loops are identified and corrected, followed byreview and modification of the structures to generate Het states using "Epik" at any specified pH and finally structure minimization using "OPLS3e force field" was performed to obtain lower energy conformation[16]. The minimised structure was used for virtual screening and molecular docking studies.

2.4 Ligand preparation

The 22 compounds identified by(Krishnan, et al., 2023), was taken for the ligand preparation[13]. The 3D coordinates of the ligands were obtained by PubChem (Table 1). These ligands were prepared using LigPrep module, that generate 3D structure with low energy under OPLS3e forcefield.

2.5 Sitemap generation

The sitemaps were generated to evaluate the potential receptor binding sites. The top ranked five sites were taken and incorporated.

Table 1. Shows the list of compounds along with their pubchem ID (CID's).

S.No	Compounds list	CID's
1	Phenol, 3-methyl-	342
2	Phenol, 2-methoxy-	460
3	Phenol	996
4	Propanoic acid, 2-methyl-, anhydride	7346
5	2-Furanmethanol	7361
6	Eicosane	8222
7	Hexadecane, 1-iodo-	11007
8	2-Cyclopenten-1-one	13588
9	2-Cyclopenten-1-one, 2-methyl-	14266
10	2-Cyclopenten-1-one, 2,3-dimethyl-	14270
11	Ethanone, 1-(2-furanyl)-	14505
12	2-Cyclopenten-1-one, 3-methyl-	17691
13	Dodecane, 2,6,11-trimethyl-	35768
14	2,3-Dimethoxyphenol	78828
15	Phenol, 2,4-bis(1,1-dimethylethyl)-, phosphite (3:1)	91601
16	2-Cyclopenten-1-one, 3,4-dimethyl-	121710
17	Undecane, 4,7-dimethyl-	519389
18	1-Hydroxy-2-butanone	521300
19	Nonane, 5-(2-methylpropyl)-	545936
20	4-Hepten-3-one, 4-methyl-	5362893
21	Tetracosane, 1-iodo-	11282694
22	Octacosane, 1-iodo-	12696144

2.6 Receptor grid generation

The receptor grid was generated on the highly ranked binding site with large volume. The grid box was kept on the center of the binding site.

2.7 Molecular docking

The docking studies were performed using Maestro version v12.5. The ligands were docked using Glide Xtra Precision (XP). The output files generated after grid generation and ligand preparation was used as input file. The docking results were obtained based on the Glide score and protein-ligand interactions.

2.8 Molecular dynamics (MD) simulations

The MD simulations was conducted for the protein-compound complex which showed high binding affinity in the docking study. This is done to estimate the stability and configuration of the complex. The complexes were subjected to 10 nanosecond (ns) MD simulation using OPLS4 forcefield. The simulation was done using Desmond program, where the conditions are build using system builder by the addition of 8 Na^+ ions and 0.15 M NaCl under an orthorhombic box shape with 10x10x10 Å in SPC solvent model. After the simulation, the interaction diagrams along with the other parameters were evaluated[17].

Table 2. Compounds list with their volume and SASA

Compound No.	Volume	SASA
1	461.527	309.830
2	482.075	318.588
3	400.514	276.833
4	663.275	417.726
5	392.720	271.582
6	1337.493	786.322
7	1165.196	694.078
8	376.631	265.090
9	436.838	297.355
10	493.011	324.629
11	434.018	292.188
12	438.879	299.064
13	1009.072	595.304
14	554.783	349.673
15	2084.699	975.791
16	488.379	319.835
17	892.578	535.560
18	403.537	279.662
19	889.046	531.659
20	596.111	383.095
21	1650.982	960.007
22	1894.490	1092.768

2.9 Molecular dynamics (MD) simulations

The MD simulations was conducted for the protein-compound complex which showed high binding affinity in the docking study. This is done to estimate the stability and configuration of the complex. The complexes were subjected to 10 nanosecond (ns) MD simulation using OPLS4 forcefield. The simulation was done using Desmond program, where the conditions are build using system builder by the addition of 8 Na^+ ions and 0.15 M NaCl under an orthorhombic box shape with 10x10x10 Å in SPC solvent model. After the simulation, the interaction diagrams along with the other parameters were evaluated[17].

2.10 Molecular dynamics (MD) simulations

The MD simulations was conducted for the protein-compound complex which showed high binding affinity in the docking study. This is done to estimate the stability and configuration of the complex. The complexes were subjected to 10 nanosecond (ns) MD simulation using

Table 3. Shows the docking score of PhcA and PhcR proteins with the listed of compounds

Compound no.	Docking score (KCAL/MOL) with PhcA	Docking score (KCAL/MOL) with PhcR
1	-2.793	-3.223
2	-4.341	-3.679
3	-4.588	-3.088
4	-3.953	-2.484
5	-1.732	-3.646
6	-4.03	-1.598
7	-3.318	-3.441
8	-1.996	-2.554
9	-2.886	-2.919
10	-2.893	-3.236
11	-3.243	-3.054
12	-3.318	-2.677
13	-2.84	-1.703
14	-0.805	-3.934
15	-4.352	-2.55
16	-1.25	-2.544
17	-3.695	-1.807

Compound no.	Docking score (KCAL/MOL) with PhcA	Docking score (KCAL/MOL) with PhcR
18	-0.852	-2.463
19	-3.163	-3.343
20	-3.219	-2.506
21	-2.959	-3.123
22	-1.363	-2.819

OPLS4 forcefield. The simulation was done using Desmond program, where the conditions are build using system builder by the addition of 8 Na^+ ions and 0.15 M NaCl under an orthorhombic box shape with 10x10x10 Å in SPC solvent model. After the simulation, the interaction diagrams along with the other parameters were evaluated[17].

3. RESULTS

3.1.Model generation and validation

The 3D structure models of the PhcA and PhcR were predicted using Itasser (Fig. 1a and 1c) and Alphafold servers (Fig. 2a and 2c) and modelled structures were evaluated using PROCHECK online server. Models obtained from Itasser resulted about 92.5 % residues were present in most favoured region, followed by 4.2 % residues in generously allowed regions and 3.3 % in disallowed regions for PhcA model (Fig. 1b). For PhcR model, about 95.4 % residues were present in most favoured region, followed by 1.2 % residues in generously allowed regions and 3.4 % in disallowed regions (Fig. 1d). Similarly, the models obtained from Alphafold resulted in, about 96.8 % residues in the most favoured region, followed by 1.6 % residues in generously allowed regions and 1.6 % in disallowed regions for PhcA model (Fig. 2b).

While, about 99.4 % residues were present in most favoured region, followed by 0.3 % residues in generously allowed regions and 0.3 % in disallowed regions for PhcR model (Fig. 2d). We could observe that the protein structures modelled using Alphafoldgave good residue coverage of about 98.4% for PhcA and 99.7% for PhcR protein structures and hence considered for further studies.

3.2.Compounds for molecular docking

The 22 compounds thus selected were downloaded from PubChem and prepared by generating 3D structure using Schrodinger LigPrep module. Energy minimization was obtained using OPLS3e forcefield and taken further for molecular docking. Table 1 shows the compounds along with their compound ID's. We can observe that the major of the compounds obtained in the AQ were phenolics. This phenol groups are the resultant of the breakage of lignin in the biomass.

3.3.Molecular docking of PhcA and PhcR protein with the compounds

The binding sites of the protein was predicted using sitemap generation. The highly ranked site was taken for grid generation. The grid box was generated at the centroid of the binding site with the dimension of PhcA (x=11.13, y=3.24, z=-15.3) and PhcR (x=12.59, y=-1.87, z=-17.82). The compounds were docked using Glide Xtra Precision (XP) for higher accuracy and precision.

3.4.Solvent accessible surface area (SASA)

The protein stability and interaction with its compound is calculated using solvent accessible surface area (SASA). Table 2 shows the volume occupies by the compound along with their SASA.

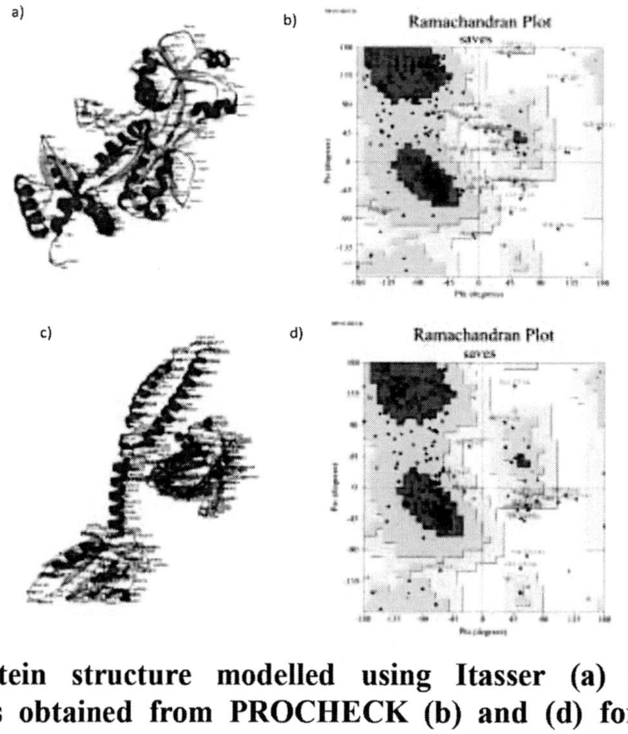

Figure 1. The protein structure modelled using Itasser (a) and (c) and the Ramachandran plots obtained from PROCHECK (b) and (d) for PhcA and PhcR respectively.

3.5.Docking score and protein-compound interaction

From the docking results we can observe that compound 3 and compound 14 gives the highest score of -4.588 and -3.934 KCal/mol for PhcA and PhcR proteins respectively. (Table 3). The docked conformation along with the protein – compound interaction diagrams are represented in Figure 3. From the interactions we can see that the -OH (hydroxy) group was responsible for the strong hydrogen bonding with the protein binding site.

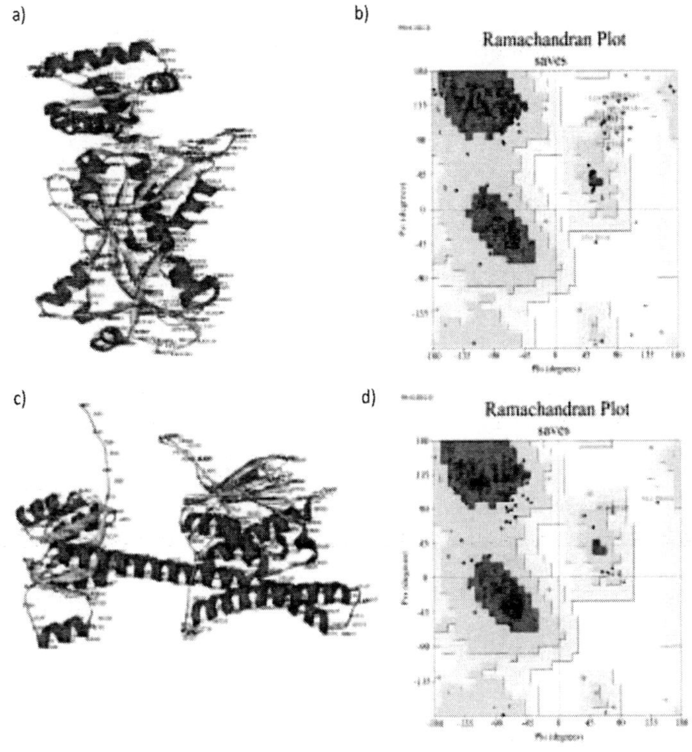

Figure 2. The protein structure modelled using Alphafold (a) and (c) and the Ramachandran plots obtained from PROCHECK (b) and (d) for PhcA and PhcR respectively.

3.6.Molecular dynamics analysis of the trajectory

Molecular dynamics (MS) provides an insight into the actual interaction that is taking place within the system. The PhcA-Phenol and PhcR-2,3 dimethoxy phenol complexes were run for 10 nanoseconds (ns). The measure of change in displacement with reference to time gives the Root Mean Square Deviation (RMSD). Figure 4a.and 5a. shows the RMSD plot of the PhcA and PhcR protein along with their compound. The complexes (PhcA- Phenol), (PhcR-2,3Dimethoxy phenol) were stabilised at 6 ns and 2 ns respectively and the complex was stable over the time frame. The protein compound contact along with the measure of its residue interaction is shown in Figures 4b and 5b. The darker shade of orange indicates more contact of the residue with the ligand. Figures 4c and 5c shows the parameters of the compound namely RMSD, radius of gyration, SASA and PSA. We could observe that the PhcR-2,3 Dimethoxy phenol complex is more stable than PhcA-Phenol complex. The 2,3-dimethoxy phenol majorly interacts with the ARG333 and GLU 152 residues, which can be seen form Figure 5d, while there is only polar solvent interaction seen in case of phenol (Figure 4d). The movie files of the trajectories are attached as separate movie files (mov1_PhcA and mov2_PhcR) for better visualisation.

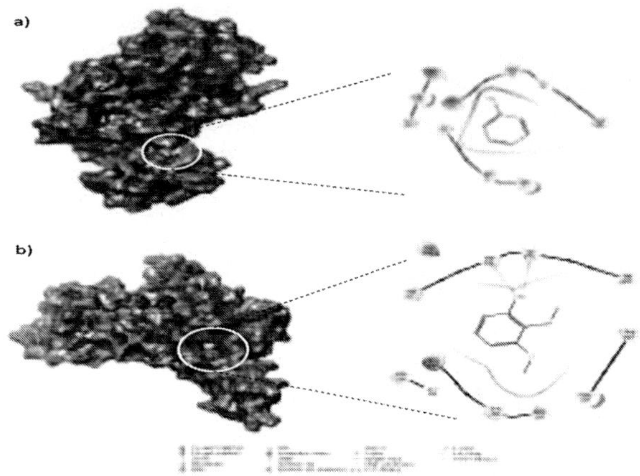

Figure 3. Docked conformation of protein with the compound along with their compound interaction diagram is shown, for a). PhcA and b).PhcR respectively.

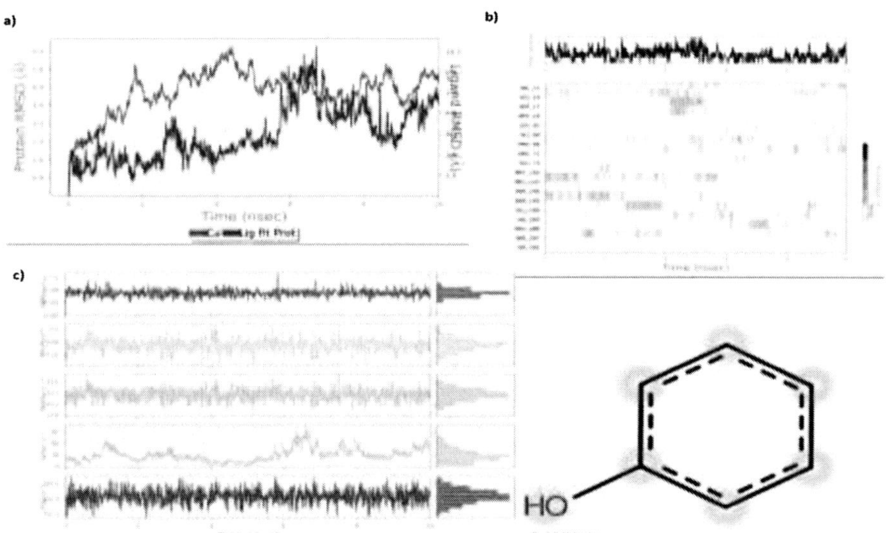

Figure 4.Protein-compound interaction of PhcA-Phenol complex showing a).RMSD, b).Protein-compound contact c).ligand RMSD, radius of gyration (rGyr), SASA and polar surface area (PSA) plots and d). Protein-compound complex interaction.

4. DISCUSSION

The Centre for Science and Environment (CSE) has recently banned 116 pesticides in India due to its proven harmful effects on human health[18]. The use of antimicrobial compounds with no toxicity to human health would be a better alternative. The use of antibacterial agent for various plant extracts have been used previously. Previous study have shown use of gallic acid against *R. solanacearum,* which showed potential biofilm inhibition[19]. Another study have shown that the extracts of *Eichhorina crassipes, Mimosa diplotricha, Lantana camara,*

and *Prosopis juliflora* were effective against *R. solanacearum.* They showed that combined treatment of these extracts drastically reduced the disease condition of the plants[20].

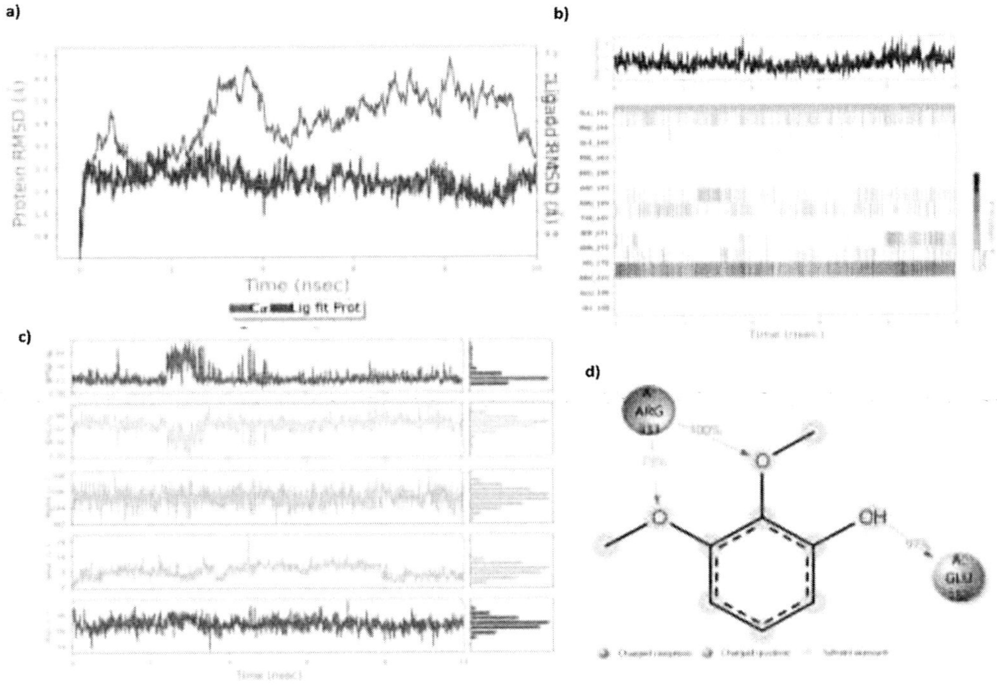

Figure 5.Protein-compound interaction of PhcR-2,3 Dimethoxy phenol complex showing a). RMSD, b).Protein-compound contact c).ligand RMSD, radius of gyration (rGyr), SASA and polar surface area (PSA) plots and d). Protein-compound complex interaction. c). and d). Protein-complex interaction.

Also, reports have showed that use of salicylic acid and leaf powder of *A. nilgiricum* increased the immunity of the plant against *R. solanacearum,* which also improved seed germination[21]. With the ongoing research, the use of nanoparticle based antimicrobial agent is of foreseen interest. Study has shown use of silver – selenium nanoparticles (Ag – Se NPs) as a best treatment in reducing the *R. Solanacearum*[22]. The recent research focusses on using the plant extracts to combat the pathogenicity. A similar study has used silver nano particles using *Rubus fruticosus*extracts to combat the Phyto pests[23]. A study by Opondo 2023, has focussed on extraction of *Tithonia diversifolia* leaves and flowers and showed it efficacy as antibacterial agent against the *R. solanacearum*.

In this context, we have investigated on the use of waste agricultural residue, wheat straw as our choice of biomass. The present work mainly deals with screening the compounds of aqueous phase of wheat straw (WS AQ), as an effective anti-infective agent. The crystal protein structures of PhcA and PhcR was not available in the protein data base (PDB) hence

they were modelled using online servers. Itasser and Alphafold servers were used to get the maximum coverage structure build. Both the modelled structures were then evaluated using Ramachandran plot to get the residue coverage, which in turn tells us the quality of the protein structure modelled. We could see that the protein structures modelled using Alphafold gave good residue coverage of about 98.4% for PhcA and 99.7% for PhcR protein structures.

The PhcA and PhcR structures, were then energy minimised and prepared for docking studies. Similarly, the structures of the 22 compounds that were reported in the WS AQ was download using PubChem. Compound preparation was done using LigPrep module. Then prepared compound and protein structures were then docked in xtra precision (XP) mode. Compounds phenol and 2,3 Dimethoxy phenol gave the maximum binding affinity for the protein PhcA and PhcR respectively. The docked confirmation was then system built by addition of solvent and ions (Na^+, Cl^-), and molecular dynamics simulation was performed. Both the complexes, PhcA-Phenol and PhcR-2,3 Dimethoxy phenol showed good and stable binding over the time frame. Also, 2,3 Dimethoxy phenol showed quick stability and more interaction with the protein (PhcR) than the phenol compound. Hence, from this study we can conclude that the phenolic compound (2,3 Dimethoxy phenol) shows binding and conformational stability over the time period for the modelled protein PhcR.

From our study results, we could see that the compounds present in the aqueous phase could potentially be an ideal candidate that could act as pest control agent against the deadly wild phyto-pathogen *R. solanacearum.* This *in silico* approach provides us an idea to further carry out the experimentation on the field with the pure compounds.

ACKNOWLEDGEMENTS

The authorssincerely thank DST – SERB (CRG/2022/003211) for the financial support through the funded project grant. Authors gratefully acknowledged SASTRA Deemed

University for research facility support. Srividhya Krishnan sincerely thanks CSIR for her Senior Research Fellowship (SRF)- 141-74-3072/2K23/1.

LIST OF ABBREVIATIONS

1. **WS AQ** – Wheat straw aqueous phase
2. **QS** – quorum sensing
3. **MD** – Molecular dynamics
4. **SASA** – solvent accessible surface area
5. **PSA** – polar surface area

REFERENCES

1. An, Y., & Zhang, M. (2024). Advances in understanding the plant-Ralstonia solanacearum interactions: unraveling the dynamics, mechanisms, and implications for crop disease resistance. *New Crops, 1*, 100014.
2. Ahmed, W., Yang, J., Tan, Y., Munir, S., Liu, Q., Zhang, J., ...& Zhao, Z. (2022). Ralstonia solanacearum, a deadly pathogen: Revisiting the bacterial wilt biocontrol practices in tobacco and other Solanaceae. *Rhizosphere, 21*, 100479.
3. Lowe-Power, T. M., Khokhani, D., & Allen, C. (2018). How Ralstonia solanacearum exploits and thrives in the flowing plant xylem environment. *Trends in microbiology, 26*(11), 929-942.
4. Li, P., Cao, X., Zhang, L., Lv, M., & Zhang, L. H. (2022). PhcA and PhcR regulate ralsolamycin biosynthesis oppositely in Ralstonia solanacearum. *Frontiers in Plant Science, 13*, 903310.
5. Vailleau, F., & Genin, S. (2023). Ralstonia solanacearum: An arsenal of virulence strategies and prospects for resistance. *Annual Review of Phytopathology, 61*(1), 25-47.
6. Dhyani, V., & Bhaskar, T. (2019). Pyrolysis of biomass. In *Biofuels: alternative feedstocks and conversion processes for the production of liquid and gaseous biofuels* (pp. 217-244). Academic Press.
7. Pešenjanski, I., Miljković, B., & Vićević, M. (2016). Pyrolysis kinetic modelling of wheat straw from the pannonian region. *Journal of Combustion, 2016*(1), 9534063.
8. Leng, L., Yang, L., Chen, J., Hu, Y., Li, H., Li, H., ...& Huang, H. (2021). Valorization of the aqueous phase produced from wet and dry thermochemical processing biomass: A review. *Journal of Cleaner Production, 294*, 126238.
9. Abd-Elnabi, A. D., El-sawy, E. A. F., & El-Adawy, E. M. (2024). Insecticidal effects of the fast pyrolysis bio-oil against Spodoptera littoralis and Aphis gossypii insect pests. *Journal of Asia-Pacific Entomology, 27*(2), 102237.
10. Bedmutha, R., Booker, C. J., Ferrante, L., Briens, C., Berruti, F., Yeung, K. K. C., ...& Conn, K. (2011). Insecticidal and bactericidal characteristics of the bio-oil from the fast pyrolysis of coffee grounds. *Journal of Analytical and Applied Pyrolysis, 90*(2), 224-231.
11. Zhang, L., Larsson, A., Moldin, A., & Edlund, U. (2022). Comparison of lignin distribution, structure, and morphology in wheat straw and wood. *Industrial Crops and Products, 187*, 115432.
12. Krishnan, S., Sivaraman, S., Jothipandiyan, S., Venkatachalam, P., Ramiah Shanmugam, S., & Paramasivam, N. (2023). Bioprospecting of aqueous phase from pyrolysis of plant waste residues to disrupt MRSA biofilms. *Biofouling, 39*(2), 231-243.
13. Krishnan, S., Gupta, K., Sivaraman, S., Venkatachalam, P., Yennamalli, R. M., & Shanmugam, S. R. (2023). Waste to drugs: identification of pyrolysis by-products as antifungal agents against Cryptococcus neoformans. *Journal of Biomolecular Structure and Dynamics, 41*(24), 15386-15399.
14. Berman, H. M., Westbrook, J., Feng, Z., Gilliland, G., Bhat, T. N., Weissig, H., ...& Bourne, P. E. (2000). The protein data bank. *Nucleic acids research, 28*(1), 235-242.
15. Yang, J., & Zhang, Y. (2015). I-TASSER server: new development for protein structure and function predictions. *Nucleic acids research, 43*(W1), W174-W181.
16. Schrodinger. Schrodinger Release 2022-3. Mastero. NY. 2022.

17. Kumar, S., Ahmad, K., Behera, S. K., Nagrale, D. T., Chaurasia, A., Yadav, M. K., ...& Singh, H. V. (2022). Biocomputational assessment of natural compounds as a potent inhibitor to quorum sensors in Ralstonia solanacearum. *Molecules*, *27*(9), 3034.
18. Shagun. Down To Earth. 2023 [cited 2025 Jan 9]. SC directs centre to submit status report on ban of pesticides. Available from: https://www.downtoearth.org.in/agriculture/sc-directs-centre-to-submit-status-report-on-ban-of-pesticides-87302
19. Sowndarya, J., Rubini, D., Sinsinwar, S., Senthilkumar, M., Nithyanand, P., & Vadivel, V. (2020). Gallic acid an agricultural byproduct modulates the biofilm matrix exopolysaccharides of the phytopathogen Ralstonia solanacearum. *Current Microbiology*, *77*, 3339-3354.
20. Alemu, D., Lemessa, F., Wakjira, M., & Berecha, G. (2014). Inhibitory effects of some invasive alien species leaf extracts against tomato (Lycopersicon esculentum Mill.) bacterial wilt (Ralstonia solanacearum). *Archives of Phytopathology and Plant Protection*, *47*(11), 1349-1364.
21. Narasimhamurthy, K., Soumya, K., Udayashankar, A. C., Srinivas, C., & Niranjana, S. R. (2019). Elicitation of innate immunity in tomato by salicylic acid and Amomum nilgiricum against Ralstonia solanacearum. *Biocatalysis and Agricultural Biotechnology*, *22*, 101414.
22. Attia, M. S., Salem, S. S., Elakraa, A. A., Abdel-Maksoud, M. A., Malik, A., Kiani, B. H., ...& El-Sayyad, G. S. (2024). Promising antagonistic effect of bimetallic silver-selenium nanoparticles against Ralstonia solanacearum-causing wilt disease in eggplant (Solanum melongena L.). *Physiological and Molecular Plant Pathology*, *133*, 102369.
23. Khan, A., Ahmad, N., Fazal, H., Ali, M., Akbar, F., Khan, I., ...& Jalal, A. (2024). Biogenic synthesis of silver nanoparticles using Rubus fruticosus extract and their antibacterial efficacy against Erwinia caratovora and Ralstonia solanacearum phytopathogens. *RSC advances*, *14*(9), 5754-5763.

Review Paper

INDIAN HERBAL FORMULATION FOR REPIGMENTATION OF VITILIGO

K. Mythili[1], Dr. S.Prabakar[2], K.Varshini[3], T.Ashwanthika[4], S.L.Tharun[5]

Assistant Professor[1], Department of Biomedical Engineering, Sona college of Technology, Salem, Tamil Nadu, India
Head of the Department[2], Department of Biomedical Engineering, Sona college of Technology, Salem, Tamil Nadu, India
Final year students[3,4,5], Department of Biomedical Engineering, Sona college of Technology, Salem, Tamil Nadu, India

ABSTRACT

Non-Segmental Vitiligo (NSV) is a common autoimmune skin condition that causes patches of skin that are depigmented because of a progressive loss of melanocytes. Repigmentation, as a vitiligo treatment method, focuses on restoring the original skin colour. Current therapies like topical steroids, narrowband UVB, and surgical interventions are commonly used but can carry risks such as skin thinning, irritation, or long-term health concerns. Our goal is to develop an herbal-based solution for repigmentation isinnovative, aiming toprovide a treatment thatavoids these side effects. This study analyses the potential of *Psoralea corylifolia* and *Moringa oleifera* extracts as a possible natural therapy solution for non-segmental vitiligo. Both plants are rich in antioxidant compounds and antimicrobial properties which help neutralize free radicals and reduce oxidative stress, a factor often implicated in skin depigmentation conditions in vitiligo. Bakuchi is high in psoralen, when the skin is treated with psoralen and subsequently exposed to UVA light, photochemical reactions occur, where psoralen forms covalent bonds with DNA, leading to cross-links between DNA strands. This process stimulates melanocyte activity, induces DNA repair mechanisms, and promotes the migration of melanocytes from hair follicles into depigmented areas. Moringa oleifera, recognized for its high antioxidant and skin-repairing properties, contains tyrosinase, an essential enzyme for melanin synthesis. The tyrosinase found in Moringa leaves can enhance melanin production and support melanocyte regeneration. The results demonstrate that this natural formulation addresses both the underlying pathophysiology and cosmetic elements of vitiligo, offering a treatment for those who suffer from the disorder. Bakuchi and Moringa extracts work together to enhance melanocyte activity, which makes this treatment a good substitute for traditional therapies.

KEYWORDS: *Repigmentation, Psoraleacorylifolia, Moringaoleifera, Psoralen, Tyrosinase.*

1. INTRODUCTION

Vitiligo is a non-familial depigmenting skin condition characterized by the selective disappearance of melanocytes, thus resulting in sharply demarcated, chalky white macules that lack pigment[1]. Such characteristic lesions can be highly psychologically disturbing and significantly affect day-to-day life, countering the notion that it is only a cosmetic affliction[6]. Affecting around 0.1%–2% of the global population, vitiligo primarily appears on exposed areas such as the face, hands, and other regions with skin pigmentation, although it can affect the entire body[5]. Advances in our understanding of vitiligo have classified it as an autoimmune disease associated with genetic susceptibility and environmental triggers[1]. There seems to be 80% genetic contribution, leaving behind 20% that contributes to the development of vitiligo through environmental triggers[5].In otherwords, oxidative stress ties this disease to melanocytes damage and to subsequent complexity in the immune system in the form of immune cytokines and exosome secretion, which keeps leading tissue to destruction. As a result, cytotoxic CD8+ T cells and interferon-γ (IFN-γ) are generated in a feedback loop, which repeatedly recruits T-cells to affected areas[6].

In 2011, an international consensus categorized vitiligo into two major types: nonsegmental vitiligo (NSV) and segmental vitiligo (SV). NSV encompasses subtypes such as acrofacial, mucosal, and generalized vitiligo, and exhibits symmetrical patterns across the body [1]. Conversely, SVgenerally involves one side and can have an earlier onset age, with a more steady progression[6]. It is important to be able to distinguish SV from NSV because this helps in prognostics and the general spread seems usually localized for SV whilst spread is over larger fields with NSV[6]. Vitiligo's complex etiologic involves an interplay of genetic, environmental, and metabolic factors, with melanocytes in patients exhibiting heightened sensitivity to oxidative stress, leading to immune cell activation and prolonged inflammation[5].

Recent research on CD8 tissue-resident memory Tcells suggests that these cells are involved in maintaining the disease and in relapses, and the chronic nature of vitiligo opens up possibilities for targeted therapeutic interventions that might disrupt this feedback loop[6].Present treatment approaches seek to control vitiligo by reducing or arresting its progression and restoring pigmentation in the skin wherever possible. Although no cure is available, treatments such as topical steroids may help limit the spread of lesions in nonsegmental vitiligo, especially when less than 10 % of the body surface area is affected[1]. Recent advances in understanding the mechanisms of the disease open up new avenues for better therapeutic approaches and promise more effective, tailored interventions to reduce the physical and psychological impacts of vitiligo.

1.1.Pathophysiology

Vitiligo is a multifactorial disorder in which the functional melanocytes are selectively lost, resulting in areas of depigmentation on the skin. Althoughmany mechanisms have been proposed over the years, the current consensus isthat it is an autoimmune condition, and that innate and adaptive immune responses participate in the progressive loss of melanocytes. Key mechanisms are genetic predispositions, oxidative stress, liberation of inflammatory mediators and cellular detachment processes-all of which go to determine the melanocyte destruction witnessed in patientspresenting with the disease of vitiligo[6].It has been estimated that about 80 % is contributed to by genetic sources while environmental impacts contribute to another 20 %. The merger or combined theory postulated that one mechanism can dovetail to work with any other thus leading to eventual depigmentation[6,20]. One of the primary triggers in vitiligo is increased melanocyte sensitivity to oxidative stress. Melanocytes produce DAMPs upon oxidative insult, which stimulates the innate immune system. For example, proteins such as heat shock protein Hsp70i facilitate the ingestion and presentation of melanocyte antigens, which in turn activate DCs and further promote immunity[26]. In this process, DCs play a central role by priming T cells, which eventually breaks immune tolerance against melanocytes and leads to their destruction. In addition, Hsp70i has been shown to enhance DC cytotoxicity, which could accelerate melanocyte damage in the context of cellular stress[26].

This immune attack on melanocytes includes both cellular and humoral components, centrally mediated by CD8+ cytotoxic T cells. Activated CD8+ T cells produce interferon-γ (IFN-γ), a cytokine that stimulates keratinocytes to release chemokines CXCL9 and CXCL10, which then recruit more autoreactive CD8+ T cells into the skin. This constitutes a self-sustaining positive feedback loop in which immune cell recruitment leads to continuing melanocyte destruction[26]. In addition to T-cell activity, tissue-resident memory T cells contribute to long-term maintenance and potential relapse of vitiligo by mediating sustained immune activity around melanocyte cells[5]. Once a bystander, the latest evidence has established the involvement of the keratinocytes in the pathogenesis of vitiligo.These skincells critically play a role in producing IFN-γ-induced chemokines, CXCL9 and CXCL10, which are needed for the recruitmentof T cells into affected skin regions[26]. The role of keratinocytes in combination with continuous infiltration by immune cells depicts how complex and chronic the process is of vitiligo. This multi-dimensional pathophysiologic framework highlights a need for treatments involving both immune modulation and antioxidant therapy to provide effective and long-lasting improvement in vitiligo management.

1.2. Prevalence

Vitiligo is the most prevalent depigmenting skin condition, affecting 0.5-2 percent of the global population in both adults and children. In 1977, one of the earliest and largest epidemiological studies was conducted on the Isle of Bornholm, Denmark, where vitiligo was found to affect 0.38 percent of the population[6]. South Asian populations including Mexican Americans. Of all casesof skin cancer recorded in the period 2003–2007 at hospital Kuala Lumpur, Malaysia only about 2.2% were new. It is found not to be restricted by age or sex as seen among patients from diverse ethnic and racial backgrounds with differing configurations related to their pigmentary nature under examination. Vitiligo has been found in all ethnic races particularly from Malaysia. Vitiligo affects people of all ethnicity and skin types equally[18]. Males and females are equally impacted, yet women and girls seek counsel more frequently probably because the negative social influence than for men and boys. Non segmental vitiligo can occur in all ages, but approximately fiftypercent of thepatientsdevelop it before 20 years old; most commonly at around age ten. About 25% of patients with vitiligo get the condition before they are 10 years old, almost half develop disease before age at just over a week and upwards to about seven or eight out of ten between periods into their third decade. Populations with bimodal distribution of the age of onset or more than one peak were considered to have an intermediate effect size as far reported. Age of onset for segmental vitiligo: Segmental occurs more commonly at a younger age than non-segmental; 87% developed before the age of 30 years and in some series, over one third (41.3%) with presentation by ten-years-old[6-16]. Individuals with dark skin (African and Asian) are more stigmatized, discriminated against and may experience high psychosocial disturbance in which they live the spectrum of emotional stress forged through human component (Figure 1)[19].

2. TYPES OF VITILIGO

Two broad classifications for vitiligo exist, non-segmental vitiligo (NSV) and segmental vitiligo (SV), which together characterize the extent of vitiligo spread. The majority form of the disease, or NSV, also is described as a bilateral or generalized condition, predominantly presenting on symmetric body parts-often including areas on the face, neck, hands, and feet. NSV encompasses various subtypes: each different subcategory is delineated according to the appearance and arrangement of depigmentation. Generalized vitiligo features widespread, scattered white patches across various areas, while acrofacial vitiligo is specific to the facial orifices and distal fingers. Universal vitiligo (Figure 2), a rare form, involves over 80% of the body's surface area[1]. Additionally, mixed vitiligo combines segmental vitiligo with any non-segmental form, often resulting in complex depigmentation patterns[16].

Mucosal vitiligo is another variety of NSV, in which mucous membranes are affected, mostly involving oral and genital mucous membranes. Focal vitiligo is a precursor to generalized forms and typically appears as isolated patches within a small area, and this pattern remains stable over time. Unique regional types of vitiligo include "lip-tip" vitiligo, where depigmentation occurs around the lips, fingertips, and toes primarily in individuals from Southern Asia[12]. Certain chemicals can also cause vitiligo-like depigmentation. It typically begins at the site of contact and spreads to other non-exposed body areas, like what occurs in other NSV forms[1].

Segmental vitiligo (SV), also known as unilateral vitiligo, generally occurs in younger patients and is less frequent than NSV. This is characterized by block-like or linear depigmented patches on one side of the body, and typically, does not cross the midline. SV often rapidly progresses over six months to two years and then stabilizes, often with the appearance of leukotrichia (white hair) in the affected areas. Compared to NSV, SV is relatively insensitive to the treatment and is linked with lower autoimmune diseases as its common characteristic, which is typically more in NSV cases[1].

3.VITILIGO STANDARD TREATMENTS

Two broad classifications for vitiligo exist, non-segmental vitiligo (NSV) and segmental vitiligo (SV), which together characterize the extent of vitiligo spread. The majority form of the disease, or NSV, also is described as a bilateral or generalized condition, predominantly presenting on symmetric body parts-often including areas on the face, neck, hands, and feet. NSV encompasses various subtypes: each different subcategory is delineated according to the appearance and arrangement of depigmentation. Generalized vitiligo features widespread, scattered white patches across various areas, while acrofacial vitiligo is specific to the facial orifices and distal fingers. Universal vitiligo (Figure 2), a rare form, involves over 80% of the body's surface area[1]. Additionally, mixed vitiligo combines segmental vitiligo with any non-segmental form, often resulting in complex depigmentation patterns[16].

Mucosal vitiligo is another variety of NSV, in which mucous membranes are affected, mostly involving oral and genital mucous membranes. Focal vitiligo is a precursor to generalized forms and typically appears as isolated patches within a small area, and this pattern remains stable over time. Unique regional types of vitiligo include "lip-tip" vitiligo, where depigmentation occurs around the lips, fingertips, and toes primarily in individuals from

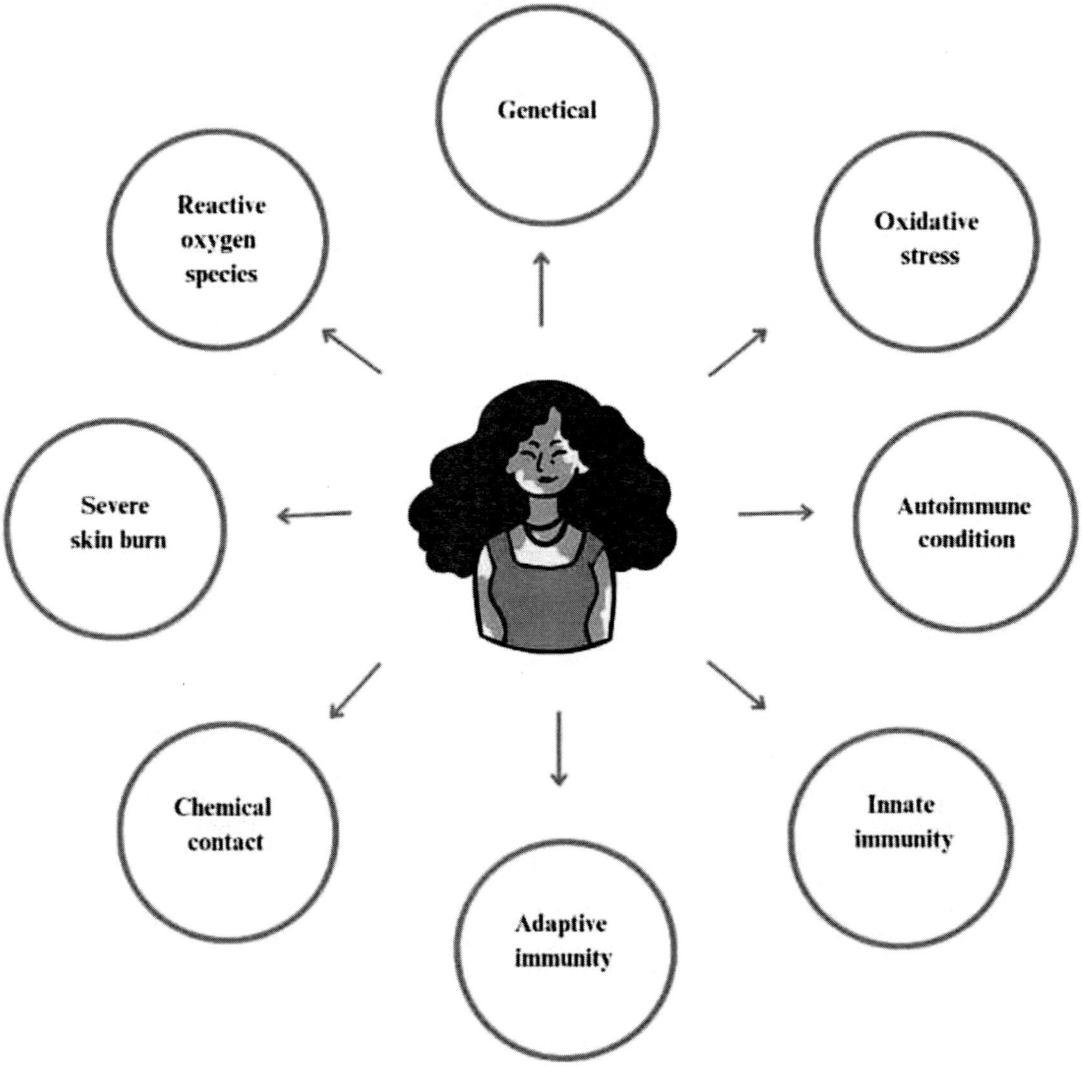

Figure 1. Causes of Vitiligo

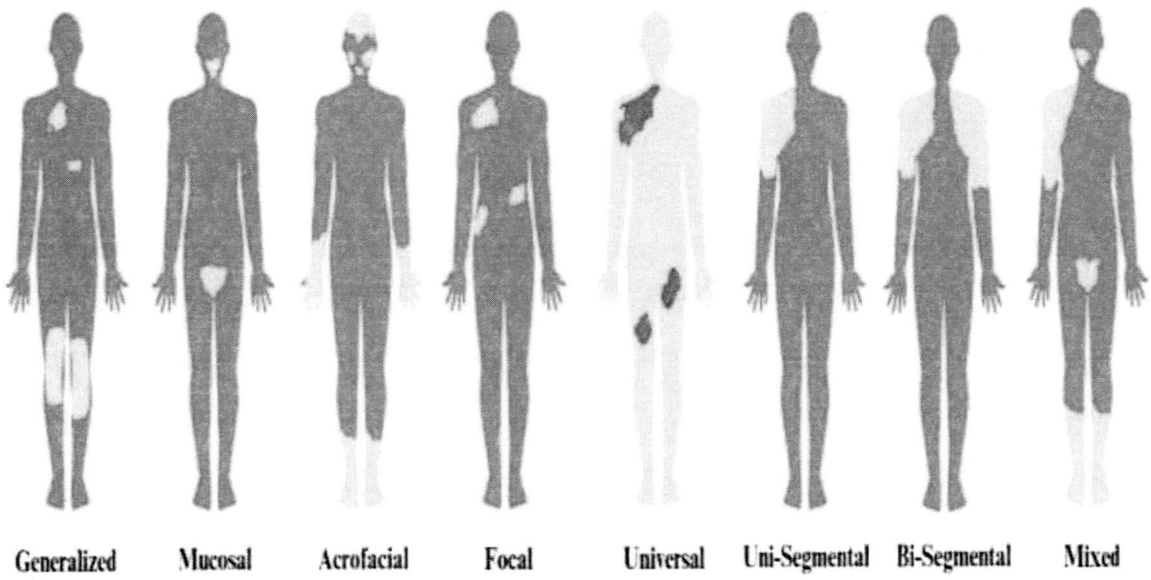

Figure 2. Types of Vitiligo

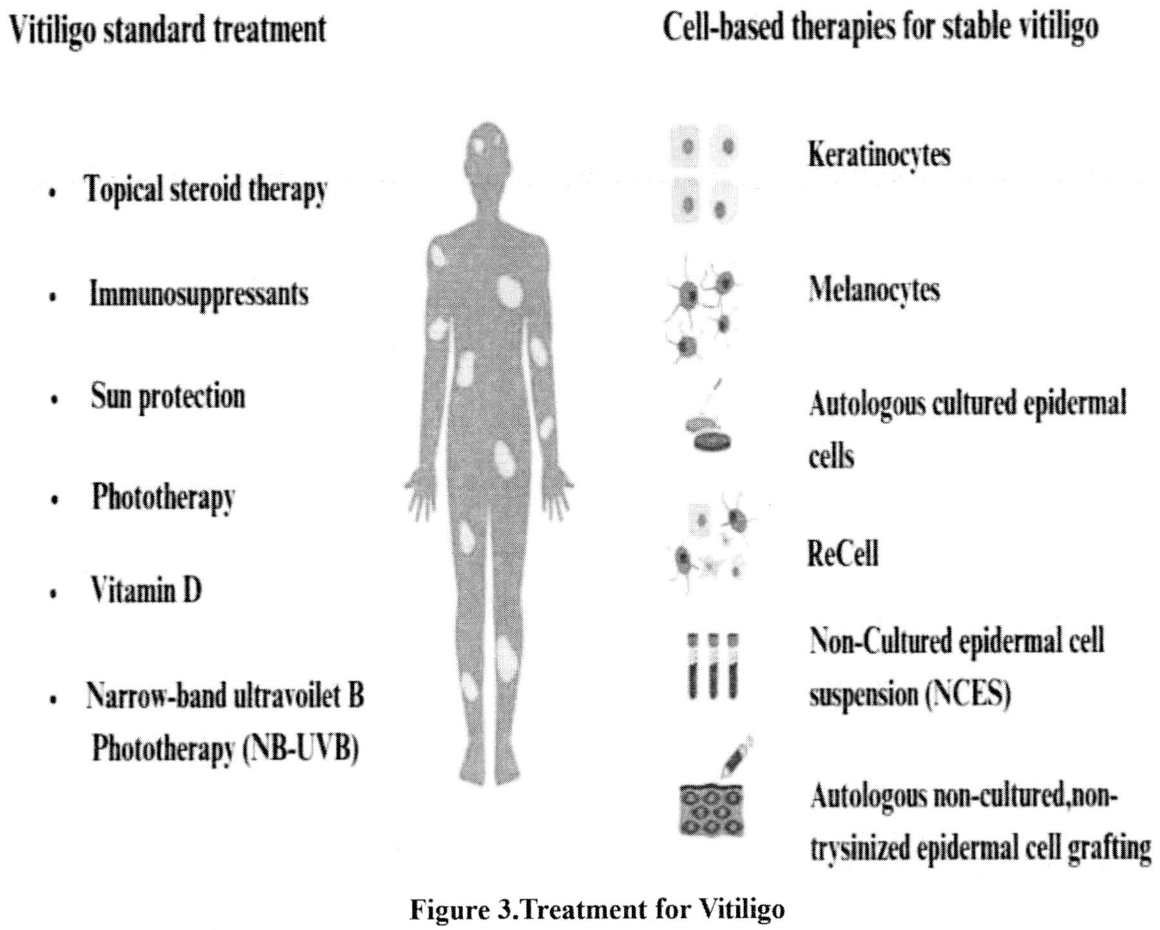

Figure 3.Treatment for Vitiligo

Southern Asia[12]. Certain chemicals can also cause vitiligo-like depigmentation. It typically begins at the site of contact and spreads to other non-exposed body areas, like what occurs in other NSV forms[1]. Segmental vitiligo (SV), also known as unilateral vitiligo, generally occurs in younger patients and is less frequent than NSV.

This is characterized by block-like or linear depigmented patches on one side of the body, and typically, does not cross the midline.SV often rapidly progresses over six months to two years and then stabilizes, often with the appearance of leukotrichia (white hair) in the affected areas.

Compared to NSV, SV is relatively insensitive to the treatment and is linked with lower autoimmune diseases as its common characteristic, which is typically more in NSV cases[1].

4. EVALUATION OF MORINGA AND BAKUCHI

Moringa oleifera is a perennial tropical deciduous arbor belonging to the family of Moringaceae; it originated in India and has been widely planted in Asia, Africa, and tropical and subtropical regions of Central America. Moringa oleifera has an important economic value apart from being an ornamental tree species because practically all of its parts, including the leaf, root and seed, may be used as food. Since it is rich in active compounds, the plant has broad pharmacological properties, including anti-oxidation, anti-inflammation, anti-diabetes, lipids lowering, anti-cancer, and anti-bacterial, among others. In this way, people have always been more interested in the plant Moringa oleifera as it is famously known as the "miracle tree"[28]. *Psoralea corylifolia* is an annual plant that has been used as a remedy in traditional Chinese and Indian medicine. The seeds of this species contain many coumarins, among which are psoralen and its isomers[2].

Psoralea corylifolia is an annual plant growing up to 50 - 90 cm tall. It bears pale purple flowers on short, condensed axillary spikes. The corolla is pale purple. Flowers bear one-seeded fruits. This plant has the most distinguishing feature in the form of minute brown glands imbedded in the surface tissue on all parts of the plant, which gives it a special and pleasant odor. The various phytochemicals present in *Psoralea corylifolia* extract include flavonoids (neobavaisoflavone, isobavachalcone, bavachalcone, bavachinin, bavachin, corylin, corylifol, corylifolin, and 6-prenylnaringenin), coumarins (psoralidin, psoralen, isopsoralen, and angelicin), and meroterpenes (bakuchiol and 3-hydroxybakuchiol)[2, 27].

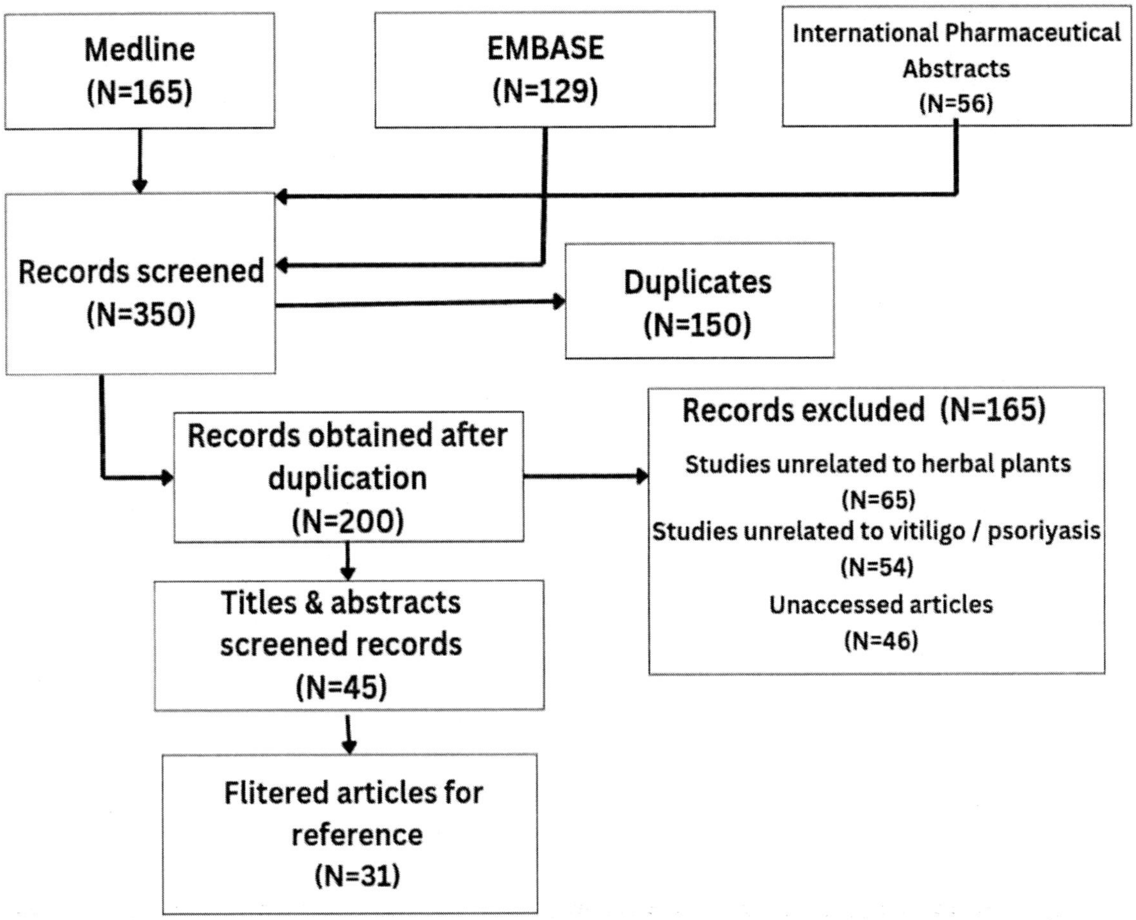

Figure 4. Asystematic review flowchart

5. CHEMICAL CONSTITUENTS IN BAKUCHI AND MORINGA

Bakuchi oil, derived from Psoralea corylifolia, was subjected to several levels of phytochemical study before its primary components were isolated and characterized. Initially, dried fruits were gathered and Soxhlet extraction with hexane was used to extract psoralen and isopsoralen. These chemicals were isolated and purified using silica gel column chromatography, then verified by TLC and GC. Other secondary metabolites such as bakuchiol, corylin, dehydroisopsoralidin, psoralidin, and isobavachalcone were also identified using a similar method.

These chemicals were confirmed using spectroscopic methods like NMR and mass spectrometry. TLC fluorodenietometry analysis was performed on the methanol extract of Bakuchi oil to standardize it and identify important phytochemicals. The quantity of psoralen, a key active component, was determined by creating a standard curve from pure psoralen and comparing it to the sample extract. The psoralen concentration was determined to be 0.083 mg/ml, and the procedure was highly precise, with a coefficient of variation of 4.3%. This systematic approach ensures the safe and effective therapeutic application of Bakuchi oil in Ayurvedic medicine[28].

Moringa oleifera is a powerhouse of bioactive compounds that contribute to its numerous medicinal properties. Flavonoids, phenols, phytosterols, and carotenoids are some of the major phytochemicals present in moringa. Melanocytes and other skin cells are protected by the powerful antioxidant effects of flavonoids like quercetin and kaempferol, which additionally decreases the inflammation and oxidative stress. Additionally, phenolic chemicals are important for protecting cells from harm, scavenging free radicals, and promoting skin health. Carotenoids such as β-carotene help to maintain healthy pigmentation and guard against UV-induced skin damage, while phytosterols, with their anti-inflammatory and cholesterol-lowering qualities, also aid in the protection of skin cells[2,27].

6. SIDE EFFECTS

Mechanism and Side Effects of Phototherapy in the Treatment of Vitiligo

Phototherapy for vitiligo should preferably use UVB-nb because it inhibits cytokine activity and prevents apoptosis in skin cells. Such treatment further activates inactive melanocytes in the hair follicle and allows these cells to proliferate and migrate into the depigmented areas in vitiligo. While having certain advantages, this mode of treatment possesses its demerits.

Principal Side Effects

1. Painful Burns: The severe burns are those occurring in patients under UVB treatment. Patients complain of skin burns in the treatment areas, mostly due to prolonged or high-intensity treatments not properly observed.
2. Itching: This relates to the sensation of itch in the treated areas that can lead to a burning sensation throughout the body.
3. Dry skin: Also known as xeroderma, it is the most common condition seen, in particular, in patients on extended or frequent phototherapy. This might lead to the skin flaking, scaling, or cracking.

Some Considered Side Effects

4. Increased Sensitivity of Skin: UVB is said to render the skin more reactive to sunlight, and subsequent exposure could cause aggravated burning even with little exposure to the sun.
5. Hyperpigmentation: Adverse regions outside the vitiligo patches sometimes become excessively pigmented due to uneven repigmentation or hyperpigmentation concurrent to the process of UVB treatment.
6. Enhanced Cancer Risk: Due to long-term ultraviolet light exposure, it can increase the risk of skin cancer, especially at high doses. However, the risk posed by UVB-nb treatment remains lower compared to PUVA (psoralen plus UVA). Prolonged treatments continue to pose this concern.

Side effects accrued can attest to the necessity for carefully intertwined therapy guidelines. The use of emollients, sunscreen, and topical creams (when taken under medical supervision) can palliate some of those side effects, further enhancing the efficacy and comfort of vitiligo treatment with UV B phototherapy[31].

7. MATERIALS AND METHOD

Moringa Extract

Due to very high antioxidant, anti-inflammatory, and immunomodulatory properties Moringa is used for the treatment of vitiligo. Fresh leaves of Moringa oleifera were collected and processed into raw, boiled, and methanolic extracts for the in vivo experiments. Compounds involved mainly are flavonoids, vitamin C and phenolics, which stimulate melanocyte activity for repigmentation. Quercetin and kaempferol are present in the leaves in a rich quantity and lower oxidative stress. For the most part, an extractive process of cold pressing or solvent extraction is applied, using ethanol or water as an extracting agent for bioactive compounds of the plant[28].

Bakuchi Extract

Psoralen and bakuchiol are the main bioactive compounds that are present in Bakuchi plant (Psoralea corylifolia) which is widely used in vitiligo treatment. Psoralen sensitizes the skin to ultraviolet light, allowing melanin production during exposure to the sun, which serves to bring color to patches of depigmented areas. Bakuchiol has properties of anti-inflammatory and antioxidant. From the plant, these compounds are isolated mainly with alcohol-based solvents, such as ethanol, for maximal effectiveness[2].

Bakuchi (Psoralea corylifolia) and Moringa are natural since they are derived from plants rather than synthesized in a lab. These plants can stimulate the cells in charge of producing melanin efficiently; that is, melanocytes, through increasing melanin repigmentation within the skin, including UV-irradiated skin. Unlike other synthetic drugs,

which may trigger side effects like skin reactions, hepatotoxicity, or immune depression, Bakuchi and its derivatives tend to provide a milder treatment approach. They would therefore be safer for long-term usage in managing Vitiligo[2,27].

8. CONCLUSION

While investigating the Indian herbal remedies for vitiligo, two botanicals, namely Psoralea corylifolia and Moringa oleifera, have shown to be effective in treating the disease. There are indications that Psooaleacorylifolia can enhance the growth and activity of melanocytes as well as the production of the pigment melanin due to its high composition of psoralen. In addition, moringa oleifera is known to possess immunomodulatory and antioxidant properties which may also help facilitate the repigmentation process. These extracts have therefore provided a more integrated method of treating vitiligo which seeks to incorporate traditional Ayurveda medicine as well as contemporary science.More clinical studies and controlled trials are also encouraged to demonstrate the optimal dosages, assess the long-term outcomes and the effectiveness in different ethnicity groups. This means integrating this approach to treatment of vitiligo may lead to development of safer, affordable and culturally appropriate options for patients with the condition in dermatology which is more organic and customized caring to the patients.

Table1. Comparative study of widely used herbs for Vitiligo treatment

TITLE	HERB	DESCRIPTION
Evaluation of a standardized extract of ginkgo biloba in vitiligo remedy	*Ginkgo biloba*	Ginkgo biloba decreases oxidative stress, reduces inflammation (IL-6), and could thereby modulate immune response invitiligo[3]
PreventionandTreatment of Vitiligo	Turmeric	This herb, Curcuma longa, stimulates skin pigmentation, has antioxidant and anti-inflammatory properties, and supports melanocyte regeneration. [20]
Efficacy and Safety of Natural Aloe Vera Gel in theTreatment of StableVitiligo	AloeVera	Promotes re-pigmentation, safe with few side effects, and has antioxidant properties along with diminished oxidative stress in stable Vitiligo [23]
PreventingVitiligowith Herbal Remedies	Neem	Neem is immunomodulatory,and strong antiseptic that can treat manyskin-related conditions such as vitiligo [12]
In Vivo Therapeutic Study of Repigmentation of Depigmented Patches in Vitiligo Disorder Isolated Tyrosinase of Moringa Oleifera and Other Extracts in Zebra Fish Embryo	*Moringa oleifera.*	Moringa oleifera is rich in bioactive compounds like phytosterols, flavonoids, and carotenoids, which promote repigmentation byenhancing melanin production [28]
Studies on the Composition and standardization of "Bakuchioil", an Ayurvedic medicinal oil prepared from *Psoraleacorylifolia* L. used in the treatment of vitiligo	*Psoralea corylifolia* (Bakuchi)	Psoralea corylifolia, or Bakuchi oil, brings repigmentation in vitiligo patients as this oil starts functioning through psoralen-a natural compound from the Psoralea corylifolia plant that makes the skin sensitive to UVA light and increases melanin pigmentation in the skin. [27]

.

REFERENCES

1. Abdel-Malek, Z. A., Jordan, C., Ho, T., Upadhyay, P. R., Fleischer, A., & Hamzavi, I. (2020). The enigma and challenges of vitiligo pathophysiology and treatment. *Pigment cell & melanoma research*, *33*(6), 778-787.
2. Abeysekera, A. M., Gunaherath, K. B., Gunawardena, A. R., & Jayaweera, C. D. (2012). Studies on the composition and standardization of" Bakuchi oil", an Ayurvedic medicinal oil prepared from Psoralea corylifolia L. used in the treatment of vitiligo. *International Journal of Research in Ayurveda & Pharmacy*, *3*(3).
3. Abu-Raghif, A. R., Ali, N. M., Farhood, I. G., FAREED, M., & HAMEED, H. B. S. (2013). Evaluation of a standardized extract of ginkgo biloba in vitiligo remedy. *Asian J Pharm Clin Res*, *6*(5), 127-130.
4. Al-Snafi, A. E. (2013). Chemical constituents and pharmacological activities of Ammi majus and Ammi visnaga. A review. *International Journal of Pharmacy and industrial research*, *3*(3), 257-265.
5. Bergqvist, C., & Ezzedine, K. (2021). Vitiligo: a focus on pathogenesis and its therapeutic implications. *The Journal of dermatology*, *48*(3), 252-270.
6. Bergqvist, C., & Ezzedine, K. (2020). Vitiligo: a review. *Dermatology*, *236*(6), 571-592.
7. Bishnoi, A., & Parsad, D. (2018). Clinical and molecular aspects of vitiligo treatments. *International journal of molecular sciences*, *19*(5), 1509.
8. Chen, J., Li, S., & Li, C. (2021). Mechanisms of melanocyte death in vitiligo. *Medicinal Research Reviews*, *41*(2), 1138-1166.
9. Chen, L., & Shen, Z. (2020). Tissue-resident memory T cells and their biological characteristics in the recurrence of inflammatory skin disorders. *Cellular & molecular immunology*, *17*(1), 64-75.
10. Costin, G. E., & Hearing, V. J. (2007). Human skin pigmentation: melanocytes modulate skin color in response to stress. *The FASEB journal*, *21*(4), 976-994.
11. Cui, T., Zhang, W., Li, S., Chen, X., Chang, Y., Yi, X., ...& Li, C. (2019). Oxidative stress–induced hmgb1 release from melanocytes: a paracrine mechanism underlying the cutaneous inflammation in vitiligo. *Journal of Investigative Dermatology*, *139*(10), 2174-2184.
12. Dwivedi, M. Preventing Vitiligo with Herbal Remedies. *Journal of Pharma and Ayurved Research*, *1*(1).
13. El Mofty, A. M., & Mofty, M. E. (1980). Vitiligo: a symptom complex. *International Journal of Dermatology*, *19*(5).
14. Ezzedine, K., & Silverberg, N. (2016). A practical approach to the diagnosis and treatment of vitiligo in children. *Pediatrics*, *138*(1).
15. Forschner, T., Buchholtz, S., & Stockfleth, E. (2007). Current state of vitiligo therapy–evidence-based analysis of the literature. *JDDG: Journal der Deutschen Dermatologischen Gesellschaft*, *5*(6), 467-475.
16. Gondalia, S. R., Chuhan, B., Rakholia, M., & Kanojiya, D. (2022). A Review on a Vitiligo and its Cure from Nature. *Gis Sci J*, *9*(5), 1383.
17. Hamzavi, I., Jain, H., McLean, D., Shapiro, J., Zeng, H., & Lui, H. (2004). Parametric modeling of narrowband UV-B phototherapy for vitiligo using a novel quantitative tool: the Vitiligo Area Scoring Index. *Archives of dermatology*, *140*(6), 677-683.
18. Le Poole, I. C., Das, P. K., Van Den Wijngaard, R. M. J. G. J., Bos, J. D., & Westerhof, W. (1993). Review of the etiopathomechanism of vitiligo: a convergence theory. *Experimental dermatology*, *2*(4), 145-153.
19. Narayanaswamy, R., & Ismail, I. S. (2018). Role of herbal medicines in vitiligo treatment-current status and future perspectives. *Asian J. Pharm. Clin. Res*, *11*, 19-23.
20. Nazar, T. Prevention and Treatment of Vitiligo.
21. Ongenae, K., Van Geel, N., & Naeyaert, J. M. (2003). Evidence for an autoimmune pathogenesis of vitiligo. *Pigment Cell Research*, *16*(2), 90-100.
22. Paarakh, P. M. (2010). Nigella sativa Linn.–A comprehensive review.
23. Padmakar, S., Kumar, G. A., Khurana, N., Kumari, S., & Pal, B. (2023). Efficacy and safety of natural Aloe Vera gel in the treatment of stable vitiligo. *Clinical Epidemiology and Global Health*, *22*, 101332.
24. Pang, Y., Wu, S., He, Y., Nian, Q., Lei, J., Yao, Y., ...& Zeng, J. (2021). Plant-derived compounds as promising therapeutics for vitiligo. *Frontiers in pharmacology*, *12*, 685116.

25. Poleo, A. G., & Rojas, J. E. (2007). Natural product in cream with anti-vitiligo therapeutic properties. *Eur. Pat. Appl.*
26. Seneschal, Julien, Katia Boniface, Andrea D'Arino, and Mauro Picardo. "An update on Vitiligo pathogenesis." Pigment cell & melanoma research 34, no. 2 (2021): 236-243.
27. Sharma, A., & Sharma, G. C. Biochemical evaluation of depigmentation in C57/BL6 mice and its treatment by Psoralea corylifolia. Linn (Bakuchi) seed oil and seed extracts.
28. Sushmitha, K., & Singh, K. G. (2023). In Vivo therapeutic study of Repigmentation of Depigmented patches in Vitiligo disorder isolated Tyrosinase of Moringa oleifera and other Extracts in Zebra Fish Embryo. *Journal of Survey in Fisheries Sciences*, *10*(1S), 6544-6553.
29. Van Den Boorn, J. G., Picavet, D. I., Van Swieten, P. F., Van Veen, H. A., Konijnenberg, D., Van Veelen, P. A., ... & Luiten, R. M. (2011). Skin-depigmenting agent monobenzone induces potent T-cell autoimmunity toward pigmented cells by tyrosinase haptenation and melanosome autophagy. *Journal of Investigative Dermatology*, *131*(6), 1240-1251.
30. Yoon, J., Sun, Y. W., & Kim, T. H. (2011). Complementary and alternative medicine for vitiligo. *Vitiligo–Management and Therapy*, 143-158.
31. Yu, R., Broady, R., Huang, Y., Wang, Y., Yu, J., Gao, M., & Zhou, Y. (2012). Transcriptome analysis reveals markers of aberrantly activated innate immunity in vitiligo lesional and non-lesional skin. *PloS one*, *7*(12), e51040.

Review Paper

BACTERIAL CELL ENCAPSULATION METHODS: A REVIEW

Akshaya M C, Anisha S, Dr. Nalinkanth V Ghone*

Department of Chemical Engineering, Sri Sivasubramaniya Nadar College of Engineering, Rajiv Gandhi Salai (OMR), Kalavakkam, Chennai -603110, Tamil Nadu, India
**Corresponding Author: Dr. Nalinkanth V Ghone*
Email id:nalinkanthv@ssn.edu.in

ABSTRACT

Probiotics play a pivotal role in enhancing digestion, boosting immunity, and preventing gastrointestinal disorders. However, their industrial applications face challenges such as reduced viability during processing, storage, and gastrointestinal transit. Encapsulation has emerged as a promising strategy to address these issues by providing a protective barrier against environmental and physiological stresses. This review examines various encapsulation techniques, including Layer-by-Layer (LbL) assembly, spray drying, freeze drying, and emulsion-based methods, which enhance microbial viability using biopolymers like alginate, chitosan, and gelatin. It consolidates insights on encapsulation mechanisms, materials, and characterization techniques while highlighting advancements, challenges, and future strategies to improve targeted delivery and industrial applications in functional foods, pharmaceuticals, and biotechnology.

KEYWORDS: *Probiotics, Encapsulation Methods, Biopolymers, Microbial Viability, Protective Agents, Functional Foods, Pharmaceutical Applications.*

1. INTRODUCTION

The human gut microbiota, a complex community of microorganisms, plays a crucial role in maintaining various physiological functions, such as digestion, nutrient absorption, immune modulation, and pathogen protection[1]. Dysbiosis, or an imbalance in this microbiota, is associated with chronic conditions like inflammatory bowel disease (IBD), obesity, diabetes, and mental health disorders. As a result, maintaining a healthy gut microbiota has become an essential aspect of human health, driving significant interest in its role in disease prevention and overall well-being. Probiotics, live beneficial microorganisms such as *Lactobacillus casei*, *Bifidobacterium breve*, and *Saccharomyces boulardii*, have shown potential in restoring gut balance and managing disorders like IBS and IBD. Despite their benefits, probiotics face challenges during industrial processing, storage, and gastrointestinal transit, which can reduce their viability and therapeutic efficacy. Encapsulation technologies, which protect probiotics from harsh conditions, offer a promising solution to these challenges. By enclosing probiotics in protective matrices, encapsulation enhances their stability and controlled release, ensuring their functionality throughout the manufacturing process and at the site of action.

Various encapsulation techniques, such as spray drying, extrusion, emulsion-based methods, and Layer-by-Layer (LbL) assembly, help maintain probiotic viability and stability during gastrointestinal transit[2]. The choice of encapsulation technique significantly impacts the effectiveness of probiotics, influencing their release profiles and functionality. Natural biopolymers like alginate, chitosan, and gelatin, along with synthetic polymers, play key roles in enhancing probiotic delivery. This review examines these encapsulation techniques, materials, and their impact on stability and release profiles, providing insights into the advancements in probiotic delivery systems and their potential to improve human health.

2. MICROENCAPSULATION: AN OVERVIEW

Microencapsulation involves enclosing micron-sized particles, liquid droplets, or gases within a protective shell to safeguard them from environmental factors like moisture, oxygen, or light. This technique enhances the stability of bioactive compounds, masks undesirable properties such as taste or odor, enables controlled or sustained release, and facilitates targeted delivery. Widely applied in pharmaceuticals, food processing, cosmetics, and agriculture, microencapsulation addresses diverse industrial needs[3]. A key advantage of microencapsulation is its ability to protect sensitive ingredients from degradation during storage or transport, thereby improving shelf life and stability. It also masks unpleasant organoleptic properties and minimizes adverse effects, as seen in encapsulated drugs like aspirin, which reduce gastrointestinal irritation. Furthermore, the technique enables controlled release by adjusting the shell's composition and thickness, allowing for sustained or targeted delivery, particularly in pharmaceuticals and nutrient delivery systems.

Recent advancements have focused on using biodegradable polymers like chitosan, alginate, and PLGA, offering biocompatibility and environmentally friendly solutions for applications such as controlled-release fertilizers and agrochemicals. As microencapsulation technology evolves, its role in enhancing bioactive delivery, sustainability, and product functionality continues to expand, demonstrating significant potential across scientific and industrial domains [8, 9].

3. ENCAPSULATION METHODS

3.1. Spray Drying

Spray drying is a widely used method for encapsulating bioactive compounds, including probiotics, by atomizing a liquid feed into fine droplets and rapidly drying them with hot air. This process transforms the feed into dry particles, making it efficient and scalable for industrial applications. The method involves preparing the feed by dispersing active ingredients in a solvent, atomizing it into droplets, drying the droplets in a hot air stream, and collecting the particles using devices like cyclone separators[4].Configurations such as co-current, counter-current, and mixed-flow systems offer flexibility for thermally sensitive materials.

Wall materials like maltodextrin, whey protein, and lipids play a key role in stabilizing and protecting probiotics. Protective agents, such as trehalose, enhance the viability of probiotics during high-temperature processing. Studies have shown that spray drying effectively encapsulates probiotics like *Lactobacilluscasei*, preserving their viability and ensuring controlled release. While spray drying is cost-effective and suitable for large-scale production, its limitations include the potential degradation of heat-sensitive materials and the need for precise parameter optimization. This technique is widely applied in food, pharmaceutical, and nutraceutical industries to improve the stability, shelf life, and bioavailability of probiotics and other bio actives.

3.2. Freeze Drying

Freeze drying, or lyophilization, is an effective method for encapsulating heat-sensitive bioactive compounds like probiotics. It involves freezing the material and removing water through sublimation under vacuum conditions, preserving sensitive compounds without exposure to high temperatures [5]. The process includes freezing the material, primary drying to remove most water, and secondary drying for residual moisture. Wall materials such as gelatin, maltodextrin, and polysaccharides protect probiotics during freezing, while cryoprotectants like trehalose prevent ice crystal damage, ensuring high probiotic survival rates. Freeze drying has proven successful for probiotics like Lactobacillus acidophilus, offering long shelf life and stability. Factors like freezing rate, temperature, and pressure influence efficiency, as ice crystal size and drying conditions affect product texture and viability. Though freeze drying offers superior preservation and survival rates, its high costs, extended processing times, and specialized equipment pose challenges. This method remains widely used in food, pharmaceutical, and biotech industries for encapsulating probiotics, enzymes, and vaccines.

3.3. Coacervation

Coacervation is a unique and versatile encapsulation method that involves the formation of a protective shell around the core material through liquid-liquid phase separation. This technique is based on the principle that certain polymers can undergo phase separation in response to environmental changes such as pH, temperature, or ionic strength[6]. During this process, the polymers self-assemble to form a gel-like phase, creating a shell around the encapsulated material. The process of coacervation begins with the formation of a polymer-rich phase, which can be triggered by changes in temperature, pH, or ionic strength. The core material, which could be a bioactive compound or probiotic, is then entrapped within this phase. The polymer-rich phase hardens to form a protective shell around the core material, thus ensuring its stability. This encapsulation method can be classified into two types: primary coacervation, where a single polymer undergoes phase separation, and complex coacervation, where two or more polymers interact to form a phase-separated shell.

Coacervation has been used successfully for encapsulating probiotics such as *Bifidobacterium bifidum*, ensuring high encapsulation efficiency and controlled release. This method has also been applied to encapsulate other bio actives like vitamins, enzymes, and antioxidants. The use of coacervation for encapsulating probiotics is particularly advantageous as it allows for high encapsulation efficiency and flexibility in the selection of materials. The efficiency of coacervation depends on several factors, including the pH, temperature, and ionic strength of the solution. The concentration of polymers and the presence of cross-linking agents also play a crucial role in determining the size, stability, and release properties of the encapsulated material. By adjusting these parameters, it is possible to tailor the properties of the encapsulated product to meet specific needs.

The advantages of coacervation include its high encapsulation efficiency, the ability to use a wide variety of materials, and the flexibility in controlling the release behaviour of the encapsulated material. However, the process can be complex and may have scalability challenges, particularly for industrial-scale applications. Despite these limitations, coacervation remains an important technique for encapsulating sensitive bioactives in food, pharmaceuticals, and nutraceuticals.

3.4. Extrusion

Extrusion is a widely used method for encapsulating bioactive compounds, such as probiotics, by forcing a mixture of the core material (e.g., probiotics, active ingredients, or drugs) and a matrix-forming material (usually polymers) through a die to form solid structures. The process involves heating or maintaining the mixture at specific temperature and pressure conditions before extrusion. The resulting extrudates may take the form of strands, granules, or pellets, which are then processed further into microcapsules or microparticles. In probiotic encapsulation, extrusion has proven effective for strains like *Lactobacillus rhamnosus* and *Bifidobacterium bifidum*, with polymer blends like polyvinyl alcohol (PVA) and polyethylene glycol (PEG) being commonly used to protect probiotics from harsh gastrointestinal conditions. For example, Schrezenmeier et al. (2018) demonstrated that encapsulating probiotics using PVA enhanced their survival rates through the digestive tract, enabling controlled release and improved therapeutic effects.

Two main types of extrusion are utilized in encapsulation: hot melt extrusion (HME) and cold extrusion. Hot Melt Extrusion (HME) involves heating the feed material until it melts and then extruding it through a die to form the desired shape or particle. Polymers such as PVA, PEG, and polylactic acid (PLA) are typically used in HME, which works well for temperature-stable materials[7].However, heat-sensitive bio actives like probiotics may lose viability under high temperatures, limiting HME's applicability for certain formulations. On the other hand, Cold Extrusion does not rely on heat and instead uses pressure to force the material through the die in its solid state, making it ideal for encapsulating temperature-sensitive compounds. Polymers like starch, gelatin, and gums are used in cold extrusion to preserve the integrity of sensitive bio actives such as probiotics.

Factors like extrusion temperature, pressure, screw speed, and feed rate significantly affect the extrusion process and the release profile of the encapsulated compounds. By adjusting these parameters, it is possible to control the release characteristics, ensuring sustained or targeted release as needed. The shape and size of the extrudates can also be customized based on the choice of die and extrusion conditions. Extrusion is widely applied in the food, pharmaceutical, and nutraceutical industries to encapsulate bioactive compounds like probiotics, vitamins, and enzymes, and its scalability makes it suitable for large-scale production. However, challenges such as achieving uniformity in extrusion and consistency in the polymer matrix persist, and ongoing advancements aim to overcome these hurdles and improve the efficiency of probiotic encapsulation using extrusion.

3.5. Emulsification

Emulsification is a technique that involves mixing two immiscible liquids, usually oil and water, to create an emulsion. In encapsulation, this method is used to disperse bioactive compounds in one phase, stabilized by the other phase, often with surfactants to prevent separation[8]. For probiotic encapsulation, emulsification has been effectively applied to strains like *Lactobacillus casei* and *Lactobacillus acidophilus*, using oil-in-water (O/W) emulsions. The probiotics are dispersed in the oil phase, stabilized by surfactants like lecithin, ensuring protection during passage through the stomach and controlled release in the intestines.

Emulsification can be classified into two types:

- **Oil-in-water (O/W) Emulsions:** Oil droplets containing the active ingredient are dispersed in a continuous water phase, stabilized by surfactants like lecithin or polysorbate. This method is suitable for encapsulating hydrophobic bio actives.
- **Water-in-oil (W/O) Emulsions:** Water droplets are dispersed in the oil phase, often used for encapsulating hydrophilic compounds.

The stability and encapsulation efficiency depend on factors like surfactant type, concentration, emulsifier choice, and homogenization conditions. High shear mixing or homogenization helps form stable emulsions by breaking the oil phase into smaller droplets. Emulsification is a simple, scalable, and versatile technique for encapsulating both hydrophilic and lipophilic compounds, providing controlled release and protection from environmental stressors. However, emulsions may face phase separation over time, affecting stability. Research into new surfactants and optimized formulations aims to address these challenges.

3.6. Layer-by-Layer (LbL) Encapsulation

Layer-by-layer (LbL) encapsulation is a technique where alternating layers of oppositely charged materials are deposited to form a protective multilayer structure around a core material, offering precise control over layer thickness, composition, and properties. It is particularly useful for encapsulating probiotics like *Lactobacillus casei* and *Bifidobacterium bifidum*, using polyelectrolytes such as chitosan and alginate, providing enhanced protection against acidic gastric conditions and bile salts, ensuring the viability and activity of probiotics upon reaching their target site in the colon. The process involves adsorbing a polyelectrolyte layer onto the core, followed by alternating layers of opposite charge. Cross-linking agents like glutaraldehyde or genipin can be used to improve stability and mechanical strength. LbL encapsulation offers flexibility in material selection, allowing the use of proteins, lipids, and nanoparticles to modify the encapsulating layers. The thickness of the layers can be controlled, influencing the release profile and stability of the encapsulated material[9]

This technique ensures controlled release of probiotics in the gastrointestinal tract, protecting them from harsh conditions like gastric pH, bile salts, and enzymatic degradation, enabling targeted delivery to the colon. Key parameters like pH, ionic strength, and polyelectrolyte concentration must be optimized to achieve high encapsulation efficiency and a controlled release profile. LbL encapsulation is versatile, suitable for a variety of bioactive materials, including proteins; enzymes, probiotics, and drugs, making it ideal for controlled release applications.

3.7 FLUIDIZED BED COATING

Fluidized bed coating is a process designed to protect sensitive materials by forming a thin, uniform layer of coating material around solid particles. This is achieved by fluidizing the particles in an air stream and spraying them with an encapsulating material. As the solvent evaporates, the material solidifies into a protective coating. Materials such as cellulose, starch, proteins, and polysaccharides are frequently used as coating agents due to their film-forming properties and stability. The technique has been employed to encapsulate probiotics for enhanced viability and controlled release. For instance, Strasser et al. (2012) used cellulose to encapsulate *Enterococcus faecium* and *Lactobacillus plantarum*. They found that adding sugars like sucrose and trehalose to the matrix improved the survival rates of the encapsulated bacteria. Similarly, Albadran et al. (2015) developed alginate–chitosan microcapsules for *Lactobacillus plantarum*, which demonstrated improved cell survival during storage and better resistance in simulated gastrointestinal conditions[10].

In addition to probiotics, fluidized bed coating has applications in the food industry. It is used to encapsulate vitamins, minerals, and food additives, enhancing product stability, flavor, and shelf life. However, the process comes with challenges, including high costs and sensitivity to elevated temperatures, which may degrade heat-labile components. Furthermore, maintaining optimal conditions such as air velocity, humidity, and spray rate is critical to preventing particle agglomeration and achieving consistent results. Thus, fluidized bed coating offers precise control over particle encapsulation, making it valuable in industries ranging from probiotics to nutritional supplements. While the method presents significant benefits in terms of functionality and scalability, optimization of parameters and cost reduction remain essential for broader implementation.

3.8 ELECTROSPINNING

Electrospinning is a technique used for encapsulating probiotics by applying an electric field (1–30 kV) to spin a dispersion of probiotics and polymers into fibers or spheres with diameters often under 1 μm. In coaxial electrospinning, two capillaries are used to create core-shell structures. The fiber structure is influenced by factors such as applied voltage, solution properties (polymer concentration, viscosity), and environmental conditions like temperature and humidity. This allows for precise control over fiber size and morphology. Studies show that electrospinning can encapsulate various probiotic strains, such as *Lactobacillus acidophilus* and *Lactobacillus rhamnosus*, with high viability rates. Encapsulation enhances their resistance to simulated gastrointestinal conditions, improving survival compared to unencapsulated controls. The method's advantages include high encapsulation efficiency, controllable fiber properties, and the use of biocompatible, biodegradable biopolymers like sodium alginate, chitosan, and starch.

However, electrospinning's high production costs, especially for large-scale manufacturing, and challenges in maintaining probiotic viability under harsher gastrointestinal conditions, such as exposure to bile salts, limit its broader application. Further optimization of the process and material selection is needed to improve its effectiveness in probiotic encapsulation [11].

4. CHARACTERIZATION TECHNIQUES FOR PROBIOTIC ENCAPSULATION

Characterizing encapsulated probiotics is vital to assess their functionality, stability, and release behavior. These techniques provide crucial information on the structural integrity of the encapsulation system, the encapsulation efficiency (EE), and the probiotics' viability during storage, transportation, and throughout the gastrointestinal tract. By evaluating these parameters, researchers can optimize encapsulation methods and materials to improve the stability and therapeutic efficacy of probiotics [10, 11].

4.1. Encapsulation Efficiency (EE)

Encapsulation efficiency (EE) measures the percentage of probiotics successfully entrapped within the carrier matrix. High EE indicates effective encapsulation, which reduces probiotic loss and enhances stability. Methods like emulsion, extrusion, and layer-by-layer approaches yield varying EE outcomes. For instance, *Lactobacillus casei* encapsulated with sodium alginate and tapioca flour using the emulsion technique achieved high EE by reducing free bacteria in the supernatant post-encapsulation. Similarly, freeze-drying methods showed 90% EE for *Lactobacillus rhamnosus ATCC 53103*, improving the controlled release of probiotics in the gastrointestinal tract.

4.2. Scanning Electron Microscopy (SEM)

Scanning Electron Microscopy (SEM) provides detailed images of the surface morphology and structural integrity of encapsulated probiotics. SEM reveals the shape, size, and uniformity of microcapsules, aiding in the assessment of the encapsulation process. For instance, *Lactobacillus acidophilus* encapsulated using gelatin and gum arabic showed agglomeration during freeze-drying, impacting particle size and capsule uniformity. These observations are essential for understanding the physical properties of probiotics and their delivery system.

4.3. Zeta Potential Analysis And Dynamic Light Scattering (DLS)

Zeta potential analysis and Dynamic Light Scattering (DLS) help assess the stability, size distribution, and surface charge of encapsulated probiotics. Zeta potential reflects the electrostatic stability of the particles, with higher values indicating stronger repulsive forces that prevent aggregation. DLS measures the hydrodynamic diameter and size distribution of encapsulated particles, providing insights into their uniformity and behavior in suspension. For example, *Lactobacillus acidophilus* microcapsules showed stable zeta potential values and uniform size distribution, essential for their effective delivery to the gastrointestinal tract.

4.4. FOURIER TRANSFORM INFRARED SPECTROSCOPY (FTIR)

FTIR analysis identifies the functional groups in the encapsulating materials and probiotics, revealing potential interactions that affect stability and functionality. For example, in Lactobacillus casei encapsulated in alginate-tapioca flour microspheres, FTIR spectra showed characteristic peaks corresponding to polysaccharide bands and amide bands, indicating interaction between the encapsulant and probiotic strain. These interactions are crucial for maintaining probiotic viability and stability, enhancing the encapsulation's effectiveness.

These characterization techniques—EE, SEM, zeta potential, DLS, and FTIR—are essential for optimizing probiotic encapsulation systems, ensuring stability, and enhancing therapeutic efficacy during storage and delivery.

5. CHALLENGES AND LIMITATIONS IN BACTERIAL CELL ENCAPSULATION

Bacterial cell encapsulation holds great potential for improving probiotic stability, viability, and therapeutic efficacy. However, challenges such as maintaining viability during harsh encapsulation processes like spray drying and freeze drying persist. Spray drying's scalability and simplicity are offset by heat-induced probiotic loss, while freeze drying, though effective, induces rehydration stress. Material selection plays a critical role in protecting probiotics and enabling controlled release in the gastrointestinal tract. Techniques like coacervation and emulsification face difficulties with uniformity and compatibility with probiotics. Scaling up production introduces complexities for methods such as extrusion and fluidized bed coating, which struggle with size uniformity and encapsulation efficiency. Additionally, high costs and technical complexity limit the adoption of advanced techniques like electrospinning and layer-by-layer (LbL) encapsulation. Storage stability is another significant issue, as encapsulated probiotics can degrade due to humidity, temperature changes, and oxygen exposure, necessitating materials with robust barrier properties. Ensuring controlled release in the GI tract remains a priority, with methods like LbL encapsulation utilizing pH-sensitive materials such as alginate and chitosan to promote colon-targeted delivery [12]. Regulatory and quality control challenges further complicate commercialization, as standardized protocols for assessing encapsulation efficiency, viability, and safety are still evolving. Addressing these limitations through ongoing research and optimization is essential to advancing probiotic therapy and enabling more effective delivery of beneficial microorganisms to the human gut.

6. Future Directions in Bacterial Cell Encapsulation

The future of bacterial cell encapsulation lies in innovative approaches that address current challenges and expand applications in probiotic delivery. Emerging techniques, such as hybrid materials like alginate-chitosan blends, nano-organized polyelectrolyte shells, and protein-polysaccharide complexes, offer superior stability and targeted release capabilities, while smart systems utilizing pH- and temperature-sensitive materials enable precise delivery in the gastrointestinal tract (GIT) [13]. Sustainability is increasingly emphasized, with eco-friendly polymers like alginate, cellulose, and pectin optimized for encapsulation, alongside green technologies such as solvent-free and energy-efficient processes to reduce environmental impact. Synbiotic encapsulation, combining probiotics with prebiotics like inulin, enhances probiotic stability, growth, and therapeutic efficacy through colonic fermentation. Advances in targeted delivery systems focus on improving specificity, using pH-responsive and enzymatic triggers for controlled release in the GIT, particularly in the colon. Cutting-edge technologies like microfluidics and nanotechnology further enable precise particle size control and scalability, paving the way for highly uniform and efficient encapsulation systems [14]. These developments promise to revolutionize probiotic therapies by enhancing stability, functionality, and environmental sustainability.

7. CONCLUSION

Bacterial cell encapsulation has emerged as a transformative technology in the field of probiotics, offering substantial improvements in the stability, viability, and targeted delivery of beneficial microorganisms. Various encapsulation methods, ranging from spray drying and freeze-drying to layer-by-layer encapsulation, have evolved to address the challenges of protecting probiotics against harsh gastrointestinal conditions. These advancements aim not only to preserve the activity of probiotics during transit but also to ensure their controlled release at the desired site for optimal therapeutic effects. Hybrid methods, the integration of smart release systems, and the exploration of biopolymer-based encapsulation have significantly enhanced the efficacy of these systems, making them more scalable and cost-effective. Additionally, synbiotic encapsulation, combining probiotics with prebiotics, holds the potential to improve both the stability and functionality of probiotics, offering new avenues for gut health. The future of bacterial cell encapsulation looks promising, with ongoing developments in materials, technologies, and sustainability playing a pivotal role in shaping the next generation of probiotic therapies. Innovations such as microfluidics, nanotechnology, and environmentally friendly biopolymers are poised to revolutionize the industrial production of probiotics, making them more accessible and efficient. As these methods continue to evolve, bacterial cell encapsulation is expected to have a profound impact on both health and industrial applications, offering a more effective, sustainable, and targeted approach to probiotic delivery.

ACKNOWLEDGEMENTS

We, the authors, Akshaya M. C and Anisha S., express our sincere gratitude to Dr. Nalinkanth V Ghone for his invaluable guidance and unwavering support throughout the preparation of this review.

LIST OF ABBREVIATIONS

1. **CFU-** Colony-Forming Unit
2. **EE-** Encapsulation Efficiency
3. **GI-** Gastrointestinal
4. **HME-** Hot Melt Extrusion
5. **IBD-** Inflammatory Bowel Disease
6. **IBS-** Irritable Bowel Syndrome
7. **LbL-** Layer-by-Layer
8. **MRS-** Man, Rogosa, and Sharpe
9. **O/W-** Oil-in-Water
10. **pH-** Potential of Hydrogen
11. **PLA-** Polylactic Acid
12. **PEG-** Polyethylene Glycol
13. **PS-** Particle Size
14. **PVA-** Polyvinyl Alcohol
15. **SEM-** Scanning Electron Microscopy
16. **W/O-** Water-in-Oil
17. **ZP-** Zeta Potential
18. **DLS-** Dynamic Light Scattering
19. **FOS-** Fructooligosaccharides
20. **WPI-** Whey Protein Isolate
21. **DWPI-** Denatured Whey Protein Isolate
22. **SA-** Sodium Alginates.

Table 1. Overview of Probiotic Encapsulation Techniques

Method	Probiotic Strain	Materials Used	Outcome	Encapsulation Efficiency (%)
Spray Drying	*L. plantarum* (MTCC 5422)	FOS, WPI, DWPI	Improved encapsulation with FOS; optimal conditions: 110±2 °C inlet temp, 55±3 °C outlet temp	70.77–72.82
	L. acidophilus (La-5)	Inulin, Hi-maize, Trehalose	High thermal protection with inlet temp 130 °C, outlet temp 76±5 °C	90.34–94.26
	Bifidobacterium bifidum	Maltodextrin and whey protein concentrate	Improved thermal stability and survival under simulated gastrointestinal conditions	~80%
Freeze Drying	*L. plantarum* (MTCC 5422)	FOS and WPI	High survivability during storage for 60 days; 2% cell viability loss with FOS+WPI compared to 9% with FOS alone	Above 98%
	L. acidophilus and *L. casei*	WPI and FOS	Encapsulation efficiency of 98% for WPI+FOS; higher survival rate in banana powder stored at 4°C compared to 25°C	WPI+FOS: 98%; WPI: 86%; FOS: 90%
	L. delbrueckii	Sodium Alginate, WPI trehalose	Improving the survival rate of probiotics at −18°C	Probiotics-WPI: 90.54%; Probiotics-SA: 84.46%
Emulsion	*Lactobacillus plantarum*	Gelatin and gum arabic	Encapsulation efficiency of 97.78% with high cell viability	97.78%
	Lactobacillus casei 431	WPI and gum arabic	Encapsulation efficiency of 84.95% and omega-3 encapsulation	84.95%
	Lactobacillus casei Shirota	Sodium alginate (0.5–3%)	Increase in encapsulation efficiency with increasing alginate concentration	86.71–95.25%
	Bifidobacterium BB-12	Sodium alginate	Not mentioned	89.71%

Method	Probiotic Strain	Materials Used	Outcome	Encapsulation Efficiency (%)
Extrusion	*Lactobacillus casei* LC01, *Lactobacillus casei* BGP93	Alginate–flaxseed or okra mucilage, alginate–botryosphaeran	High encapsulation efficiency; Improvement in the stability of encapsulated cells in refrigerated storage	Not mentioned
	Lactobacillus plantarum	Alginate–arabinoxylan	High encapsulation efficiency and resistance to gastrointestinal conditions, better than alginate beads	Not mentioned
	Bifidobacterium animalis spp. lactis BB12	Alginate–goats' milk–inulin	Inulin addition resulted in compact structure capsules; Protection of the probiotic in simulated gastrointestinal condition.	Not mentioned
Fludized Coatings	*Lactobacillus paracasei* BGP-1	Alginate–shellac	Improved microcapsule structure by reducing porosity; protection during storage at room temperature and gastrointestinal simulation	Not specified
	Lactobacillus casei subsp. paracasei LMG P-21380	Microcrystalline cellulose–trehalose–maltodextrin–vegetable wax	High rates of encapsulation efficiency; protection enhanced during encapsulation; greater stability during storage	Not specified
	Lactobacillus acidophilus La3	Stearic acid, albumin, alginate, cassava starch	Encapsulation efficiency above 90%; moist-heat resistance; suitable for incorporation in heat-treated foods	>90%

Method	Probiotic Strain	Materials Used	Outcome	Encapsulation Efficiency (%)
Electrospinning	*L. paracasei KS-199*	SA and PVA	Enhances the viability of probiotics	85.87%
	L. acidophilus 016	PVA and bacterial cellulose nanofibers (BCNF)	Indicates the possibility of BCNF-based probiotics for various commercial applications	Not specified
	Lactobacillus casei	PVA/polyethylene oxide copolymer (PEC)	Provides higher thermal stability and protection from gastrointestinal damage	Not specified
Coacervation (Simple)	*Lactobacillus plantarum*	Whey protein isolate (WPI), gum Arabic	Enhanced survival under simulated gastric conditions	83%
Coacervation (Complex)	*Lactobacillus acidophilus*	Gelatin, gum Arabic	High protection in acidic environments	89%
Layer-by-layer self-assembly	*L. rhamnosus GG*	Polyalginate, xanthan gum, chitosan	Enhancing the viability of probiotics in an SGF solution.	>6
	L. pentosus	Chitosan and sodium phytate	Enhancing the viability of probiotics in an SGF solution.	Not specified
	E. coli strain Nissle 1917	SA and chitosan	Improving probiotic viability in a gastric environment.	Not specified

Table 2: Common Encapsulation Materials And Their Properties

Material	Source	Properties
Sodium Alginate	Natural polymer derived from seaweed	High viscosity, forms a protective gel layer, enhances cell viability and encapsulation efficiency
Gelatin	Protein from animal sources	Provides dual protection in emulsions, improves encapsulation efficiency and cell viability
Gum Arabic	Polysaccharide from tree sap	Forms a stabilizing matrix, enhances protection and encapsulation efficiency
Whey Protein Isolate (WPI)	Protein derived from milk	Stabilizes emulsions, enhances encapsulation efficiency, forms protective microcapsules
Pectin	Plant-derived polysaccharide	"Forms a thick protective layer, enhances encapsulation stability"
Calcium Chloride	Inorganic salt	Facilitates external gelation in alginate-based encapsulation systems
Chitosan	Natural polymer from crustacean shells	Biodegradable, biocompatible, and forms strong polyelectrolyte complexes for enhanced stability and protection

REFERENCES

1. Yao, M., Xie, J., Du, H., McClements, D. J., Xiao, H., & Li, L. (2020). Progress in microencapsulation of probiotics: A review. *Comprehensive Reviews in Food Science and Food Safety*, *19*(2), 857-874.
2. Islam, M. A., Yun, C. H., Choi, Y. J., & Cho, C. S. (2010). Microencapsulation of live probiotic bacteria. *Journal of microbiology and biotechnology*, *20*(10), 1367-1377.
3. Rajam, R., & Subramanian, P. (2022). Encapsulation of probiotics: Past, present and future. *Beni-Suef University Journal of Basic and Applied Sciences*, *11*(1), 46.
4. Ren, Y., Jia, F., & Li, D. (2024). Ingredients, structure and reconstitution properties of instant powder foods and the potential for healthy product development: a comprehensive review. *Food & Function*.
5. Kiprono, S., Wambani, J., Rono, J., Langat, V., Shi, Z., Raballah, E., & Yang, G. (2024). Microencapsulation of Probiotics and their Applications: A Review of the Literature. *ES Food & Agroforestry*, *17*(2), 1106.
6. Azad, M. A. K., Sarker, M., Li, T., & Yin, J. (2018). Probiotic species in the modulation of gut microbiota: an overview. *BioMed research international*, *2018*(1), 9478630.
7. Yang, S., Wei, S., Wu, Y., Fang, Y., Deng, Z., Xu, J., & Zhang, H. (2024). Encapsulation techniques, action mechanisms, and evaluation models of probiotics: Recent advances and future prospects. *Food Frontiers*, *5*(3), 1212-1239.
8. Rabetafika, H. N., Razafindralambo, A., Ebenso, B., & Razafindralambo, H. L. (2023). Probiotics as antibiotic alternatives for human and animal applications. *Encyclopedia*, *3*(2), 561-581.
9. Chaturvedi, P., & Sharma, P. (2024). A Review on Microencapsulation as Method of Drug Delivery. In *BIO Web of Conferences* (Vol. 86, p. 01033). EDP Sciences.
10. silva, G.; Tavares, L.; Canhoto, J.; Amorim, M. (2020). Future perspectives on the encapsulation of bioactive compounds: A holistic approach for their efficient delivery. *Food Bioprocess Technol.* 13, 1057–1076.
11. Eratte D, Kewalramani S, Dobrev G. (2020). Microencapsulation of Lactobacillus casei 431 with omega-3 fatty acids using whey protein isolate and gum arabic. *Food Science & Nutrition*, 8(2):655-662.
12. Łętocha, A., Michalczyk, A., Miastkowska, M., & Sikora, E. (2024). Effect of Encapsulation of Lactobacillus casei in Alginate–Tapioca Flour Microspheres Coated with Different Biopolymers on the Viability of Probiotic Bacteria. *ACS Applied Materials & Interfaces*, *16*(39), 52878-52893.
13. Yang S, Zhai Y, Xie Q, et al. (2021). Applications of polylactic-co-glycolic acid (PLGA) encapsulation for bioactive compounds. Int J Biol Macromol, 165:1195-1203.
14. Agboola, B.O.; Khan, S.; Ng, W.C. (2020). A review of challenges in the production and application of microencapsulated bioactive compounds in foods. *Food Control,* 114, 107230.

Review Paper

ECO-EFFICIENT CONVERSION OF WASTE GLYCEROL TO SUCCINIC ACID: UNLOCKING SUSTAINABLE INDUSTRIAL APPLICATIONS

Dr. S.Chozhavendhan, R.Brindha*, M.Swathi

Vivekanandha College of Engineering for Women (Autonomous)
Email id :brindhar44@gmail.com

ABSTRACT

The eco-friendly conversion of waste glycerol into succinic acid offers a dual benefit of generating high-value chemicals which providing a sustainable solution to mitigate industrial waste. The underutilization of waste glycerol, a by-product from biodiesel industry, contributes to environmental problems. It explores a novel green biotechnology approach to transform waste glycerol into succinic acid, a versatile platform chemical with diverse applications in food, pharmaceuticals, and bioplastics. The fermentation process was adjusted with pH, temperature, and nutrient conditions using strains of *Escherichia coli*, enhancing the conversion efficiency of glycerol to succinic acid. The results shows increased yields of succinic acid, demonstrating the methods, economic viability and scalability. This promotes resource sustainability and waste management by turning garbage into useful resources, which is consistent with the ideas of the circular economy. This review concludes that waste valorization can integrates waste glycerol conversion into eco-friendly industrial processes.

KEYWORDS: *Waste glycerol, succinic acid, fermentation optimization, waste valorization, sustainable industrial processes, resource efficiency, environmental sustainability.*

1. INTRODUCTION

Succinic acid is a valuable platform chemical with widespread applications in industries like food, pharmaceuticals, and bioplastics due to its versatility in producing polymers, resins, and solvents. Succinic acid can be produced by fermentation as it is an intermediate of the tricarboxylic acid (TCA) and one of the end products of anaerobic metabolism[32]. The fermentation succinic acid production is carried out both with wild-type strains and with genetically modified microorganisms such as *Actinobacillus succinogenes, Basfia succiniciproducens* isolated from the rumen of ruminants. Traditionally, succinic acid was produced via petrochemical processes, but concerns over sustainability and environmental impact have spurred interest in bio-based production methods [31]. The use of renewable raw materials and carbon dioxide fixation are the main benefits when compared to the petrochemical synthesis of succinic acid. A significantly better Carbon Dioxide (CO_2) balance and a 30–40% lower energy consumption compared to conventional succinic acid obtained from fossil raw materials have led to the commercialization of bio-based succinic acid in recent years[30]. Furthermore, biomass-derived succinic acid not only reduces dependence on fossil fuels but also supports the use of renewable feedstocks, such as lignocellulosic agricultural residues and energy crops. Among potential substrates, Napier grass (*Pennisetum purpureum*) is particularly attractive due to its high cellulose content, rapid growth rate, and high annual yield, making it an ideal raw material for producing bio-based chemicals [34].

2.SCOPE AND CHALLENGES

Glycerol, a byproduct of the manufacturing of biodiesel, as a cheap carbon source for fermentation, the production of succinic acid from waste glycerol provides a commercially feasible and environmentally sustainable alternative. In addition to lowering waste disposal, this strategy offers a chance to produce succinic acid at a reasonable cost. There are still issues to be resolved, though, such as the requirement for microbial strains that are optimized to use glycerol effectively, reducing glycerol inhibition at high concentrations, reducing the production of byproducts that might compromise yield and purity, and addressing the possible deficiency of vital nutrients in glycerol. Technical challenges in preserving efficiency and cost-effectiveness also arise when the process is scaled up from laboratory to industrial levels. Resolving these issues could greatly increase the feasibility of producing succinic acid from waste glycerol[29].

3. METABOLIC PATHWAY OF MICROBES UTILIZING GLYCEROL

Microbial pathways for succinic acid production typically involve anaerobic fermentation processes that facilitate carbon flow towards succinic acid as the primary product. *Actinobacillus succinogenes* is among the most efficient succinic-acid-producing bacteria due to its ability to metabolize various sugars and glycerol. Glycerol utilization begins with its conversion to dihydroxyacetone or glycerol-3-phosphate, followed by further metabolism through glycolysis. The conversion of glucose or glycerol to succinic acid in *A. succinogenes* involves the TCA cycle, where intermediates are redirected to form succinate. In anoxic conditions, the TCA cycle operates in a reductive direction, which increases succinic acid production. The high reduction state of glycerol contributes additional reducing power, making it an effective co-substrate with glucose or other sugars to boost succinate yields. By directing metabolic flux towards succinic acid and away from byproducts like acetic and formic acids, *A. succinogenes* is able to achieve high succinic acid productivity, especially in mixed-substrate systems.

4.FERMENTATION PROCESS USING GLYCEROL

Glycerol as a carbon source in fermentation has shown promise in enhancing succinic acid yields. The process involves anaerobic fermentation in a controlled bioreactor environment where *A. succinogenes* utilizes glycerol along with glucose present in the hydrolysate. As glycerol is more reduced than glucose, it provides additional electrons, leading to a higher succinate-to-acetate (S/A) ratio, which is desirable for maximizing succinic acid production. The fermentation setup often requires optimization of glycerol and glucose concentrations, pH, temperature, and anaerobic conditions to improve yields. Previous studies have shown that when *A. succinogenes* is grown in a medium with both glycerol and glucose, succinic acid production can be significantly enhanced, as glycerol assists in maintaining a favorable redox balance within the cells. Fermentation is not necessarily carried out in an anaerobic environment. It can also be run in an aerobic way. In modern fermentation technology, aerobic conditions are maintained in a closed fermenter with submerged cultures. The culture contents are agitated with an impeller and aerated by forcing sterilized air. Table 1.1 presents earlier works on glycerol fermentation which also included the conditions being used.

5. TYPES OF FERMENTATION

Industrial fermentation processes can be divided into two main types. They are batch fermentations and continuous fermentations.

5.1. Batch fermentation process

Batch fermentation is considered a ‘closed system’. In the course of an entire fermentation, no nutrient is being added into the fermenter except for oxygen, anti-foam agent and pH controller, either in base or acidic form. In general, batch fermentation is being used to produce biomass and primary metabolite. In order to maximize biomass production, optimization of cultural conditions for both physical and chemical nutrients is highly recommended. There are a few advantages of batch fermentation such as low contamination risk, being able to run different successive phases in the same vessel and close control of genetic stability of microorganism(s). In batch fermentation mode, the systems are normally controlled by culture pH, dissolved oxygen tension and temperature. There are many researches which have applied this batch processing system to produce new products. Lee et al.[19] had successfully produced succinic acid by Anerobiosprillum succiniciproducens in batch fermentation of mixed sugar of galactose, glucose and lactose. Furthermore, Jiang et al.[32]had employed a batch fermentation system in producing succinic acid from enzymatic hydrolysate of brewer‘s yeast (EBY) by A. succinogenes NJ113 with the addition of vitamins to replace commercial yeast extract. After 48h of cultivation under batch mode, it has resulted in 46.8 g/L succinic acid concentration and yield of 68.8% being produced. Following 48 hours of batch mode cultivation, a yield of 68.8% and a succinic acid content of 46.8 g/L were obtained.

5.2. Continuous fermentation process

Continuous fermentation process is a system where the mixed culture is continuously being supplied with fresh nutrients, and volume of culture is being kept maintained by continuous removal of liquid culture at the same flow rate as the feeding rate of fresh nutrients. Hence, this method offers a continuation of growth for a long period of cell cultivation. With this feature, the designated medium could overcome the problem of limitation of substrate concentration, besides the exponential growth phase is being prolonged until an extra substrate is being exhausted on the cell. The uniqueness of this fermentation style is it operates under steady state conditions, where cell growth occurs at constant growth rate and stable environmental conditions. Factors such as pH of the culture, specific growth rate, nutrient and cell concentration, metabolites concentration and dissolved oxygen; are all kept in the continuous culture. Using this kind of fermentation method, a continuous system promises important potential advantages in reducing operating and raw material costs. Brethauer and Wyman (2009) mentioned that some continuous fermentation is now being employed for commercial ethanol production from cane sugar and corn. This is done in order to benefit from increased volumetric output, lower labor expenses, and less time spent cleaning and filling the vessel. In contrast, these systems are more susceptible to microbial contamination and need more advanced operations.

Continuous operation in succinic acid production gives more advantages over batch operation mode. Since the system is fully equipped with continuous media sterilization and aseptic technique operation, the contamination problem to the continuous system could also be eliminated. Many scientific papers and patents published today deal with the continuous fermentation system on succinic acid production. Research done by Lee et al.[33]had performed

Table 1. Bio-based production of chemical and fuels via glycerol fermentation

Product (concentration, yields*)	Substrate	Strains	Culture
1,3-Propanediol (30g/l, Y=0.61)	Crude glycerol	Clostridium acetobutylicum DG1 (psPD5)	Anaerobic (N2), continuous
Hydrogen (60mM) Ethanol (90mM, Y=0.69)	Crude glycerol	Enterobacter aerogenes HH-101	Anaerobic (N2), Batch/ continuous
Citric acid (77.4g/l, Y=0.69)	Crude glycerol	Yarrowia lypolitica NCIM 3589	Aerobic, batch
Succinic acid (4.9g/l, Y=0.59)	Impure glycerol	Anaerobiospirillum succiniciproducens ATCC 29305	Anaerobic (CO2), batch
Butanol (17g/l, Y=0.39)	Pure glycerol	Clostrodium pasteurianum DSM 525 (ATCC6013)	Anaerobic (N2), batch

anaerobic continuous culture of *Mannheimia succiniciproducens* MBEL55E in whey based medium containing Corn Steep Liquor (CSL) which had resulted in succinic acid yield of 69% and 3.9gL/h of productivity, andcontinuous culture of *Mannheimia succiniciproducens* MBEL55E in wood hydrolysate based medium and had produced succinic acid yield of 56%

and the productivity being achieved at 1.17gL/h. Besides, another research had shown another type of microbe which possesses the capabilities to produce succinic acid through the continuous system. Production of succinic acid by continuous fermentation of *Actinobacillus succinogenes* sp.130Z in an external recycle reactor for membrane cells. The output of this work was 16.4 g/L, and the productivity was 6.63 gL/h. Moreover, anaerobic continuous fermentation 10 of *Mannheimia succiniciproducens* LPK7 produced 0.38mol/mol of succinic acid and was successfully achieving productivity at 1.77gL/h[10].

6. CHALLENGES AND TECHNIQUES

The primary challenges in producing succinic acid via glycerol fermentation include:

Inhibitory Compounds: Biomass hydrolysates contain inhibitors such as acetate, which can hinder microbial growth and reduce succinic acid yield. This could be addressed by genetically modifying strains or other pretreatment techniques.

Mixed Sugar Utilization: Glucose and xylose are common sugars in hydrolysates, but *A. succinogenes* preferentially uses glucose. It can be beneficial to balance the medium's sugar content and modify microorganisms to enhance xylose utilization.

Byproduct Formation: The presence of xylose often increases the production of byproducts like acetic and formic acids. Adjusting the fermentation parameters and using glycerol can help direct metabolism toward succinic acid.

Separation and Purification: The downstream processing of succinic acid is often costly and complex. Techniques like electrodialysis, calcium precipitation, and reactive extraction are used to purify succinic acid effectively. Recent advances in membrane technology, such as bipolar membrane electrodialysis, offer more efficient separation and have been implemented to reduce costs and improve yields.

7. CONCLUSION

The bioproduction of succinic acid from lignocellulosic feedstocks—Napier grass in particular offers a viable and sustainable substitute for conventional petrochemical generation techniques. By using glycerol as a co-substrate, succinic acid yields are increased while also making use of a plentiful, renewable byproduct, which further supports environmental sustainability. This method has the dual benefit of making use of waste materials and helping to produce a useful platform chemical.

The feasibility of microbial fermentation utilizing glycerol and Napier grass hydrolysate has been effectively proven by the study, indicating promise for a variety of uses in sectors like food, medicine, and bio-based materials. This invention has the potential to significantly contribute to the shift to more environmentally friendly chemical production methods, thereby decreasing reliance on fossil fuels and the carbon footprint of industrial operations. However, further optimization is necessary for this biotechnological process to become scalable and financially successful. Enhancing microbial strains for improved performance, improving fermentation conditions, and improving downstream processing methods to increase yields and cut costs should be the main priorities. These developments have the potential to make the bioproduction of succinic acid from renewable resources a key component of a chemical industry that is more ecologically friendly and sustainable.

REFERENCES

1. Du, C., Lin, S. K. C., Koutinas, A., Wang, R., & Webb, C. (2007). Succinic acid production from wheat using a biorefining strategy. *Applied microbiology and biotechnology*, *76*, 1263-1270.
2. Dutta R., Sarkar U., Mukherjee A. Extraction of oil from Crotalaria Juncea seeds in a modified Soxhlet apparatus: physical and chemical characterization of a prospective bio-fuel. Fuel. 2014
3. Edwards, V. H. (1970). The influence of high substrate concentrations on microbial kinetics. *Biotechnology and bioengineering*, *12*(5), 679-712.
4. Fan, M., Huang, J., Yang, J., & Zhang, P. (2013). Biodiesel production by transesterification catalyzed by an efficient choline ionic liquid catalyst. *Applied Energy*, *108*, 333-339.
5. Gonzalez, R., Murarka, A., Dharmadi, Y., & Yazdani, S. S. (2008). A new model for the anaerobic fermentation of glycerol in enteric bacteria: trunk and auxiliary pathways in Escherichia coli. *Metabolic engineering*, *10*(5), 234-245.
6. Isar, J., Agarwal, L., Saran, S., & Saxena, R. K. (2006). A statistical method for enhancing the production of succinic acid from Escherichia coli under anaerobic conditions. *Bioresource technology*, *97*(13), 1443-1448.
7. Khanna, S., Jaiswal, S., Goyal, A., & Moholkar, V. S. (2012). Ultrasound enhanced bioconversion of glycerol by Clostridium pasteurianum: a mechanistic investigation. *Chemical engineering journal*, *200*, 416-425.
8. Liu, R., Liang, L., Jiang, M., Ma, J., Chen, K., Jia, H., & Ouyang, P. (2014). Effects of redox potential control on succinic acid production by engineered Escherichia coli under anaerobic conditions. *Process Biochemistry*, *49*(5), 740-744.
9. Neijssel, O. M., Teixeira de Mattos, M. J., & Tempest, D. W. (1996). Growth yield and energy distribution.
10. Okino, S., Noburyu, R., Suda, M., Jojima, T., Inui, M., & Yukawa, H. (2008). An efficient succinic acid production process in a metabolically engineered Corynebacterium glutamicum strain. *Applied microbiology and biotechnology*, *81*, 459-464.
11. Orjuela, A., Yanez, A. J., Santhanakrishnan, A., Lira, C. T., & Miller, D. J. (2012). Kinetics of mixed succinic acid/acetic acid esterification with Amberlyst 70 ion exchange resin as catalyst. *Chemical engineering journal*, *188*, 98-107.
12. Otero, J. M., Olsson, L., & Nielsen, J. (2007). Metabolic engineering of Saccharomyces cerevisiae microbial cell factories for succinic acid production. *Journal of Biotechnology*, *131*(2, suppl. 1), S205-S205.

13. Roca, C., Albuquerque, G., & Reis, M. (2010). Evolutionary engineering of Actinobacillus succinogenes for improved succinic acid production on glycerol. *Journal of Biotechnology*, (150), 373.
14. Sadhukhan, S., & Sarkar, U. (2016). Production of biodiesel from Crotalaria juncea (Sunn-Hemp) oil using catalytic trans-esterification: Process optimisation using a factorial and Box–Behnken design. *Waste and biomass valorization*, *7*, 343-355.
15. Sadhukhan, S., & Sarkar, U. (2016). Production of purified glycerol using sequential desalination and extraction of crude glycerol obtained during trans-esterification of Crotalaria juncea oil. *Energy Conversion and Management*, *118*, 450-458.
16. Saravanan, P., Pakshirajan, K., & Saha, P. (2009). Batch growth kinetics of an indigenous mixed microbial culture utilizing m-cresol as the sole carbon source. *Journal of Hazardous Materials*, *162*(1), 476-481.
17. Sdrula, N. (2010). A study using classical or membrane separation in the biodiesel process. *Desalination*, *250*(3), 1070-1072.
18. Song, H., Huh, Y. S., Lee, S. Y., Hong, W. H., & Hong, Y. K. (2007). Recovery of succinic acid produced by fermentation of a metabolically engineered Mannheimia succiniciproducens strain. *Journal of biotechnology*, *132*(4), 445-452.
19. Song, H., Jang, S. H., Park, J. M., & Lee, S. Y. (2008). Modeling of batch fermentation kinetics for succinic acid production by Mannheimia succiniciproducens. *Biochemical Engineering Journal*, *40*(1), 107-115.
20. Song, H., & Lee, S. Y. (2006). Production of succinic acid by bacterial fermentation. *Enzyme and microbial technology*, *39*(3), 352-361.
21. Tan, H. W., Aziz, A. A., & Aroua, M. K. (2013). Glycerol production and its applications as a raw material: A review. *Renewable and sustainable energy reviews*, *27*, 118-127.
22. Tan, Y., Wang, Z. X., & Marshall, K. C. (1996). Modeling substrate inhibition of microbial growth. *Biotechnology and bioengineering*, *52*(5), 602-608.
23. Vlysidis, A., Binns, M., Webb, C., & Theodoropoulos, C. (2011). Glycerol utilisation for the production of chemicals: conversion to succinic acid, a combined experimental and computational study. *Biochemical Engineering Journal*, *58*, 1-11.
24. Xi, Y. L., Chen, K. Q., Li, J., Fang, X. J., Zheng, X. Y., Sui, S. S., ... & Wei, P. (2011). Optimization of culture conditions in CO2 fixation for succinic acid production using Actinobacillus succinogenes. *Journal of Industrial Microbiology and Biotechnology*, *38*(9), 1605.
25. Yu, J., Li, Z., Ye, Q., Yang, Y., & Chen, S. (2010). Development of succinic acid production from corncob hydrolysate by Actinobacillus succinogenes. *Journal of Industrial Microbiology and Biotechnology*, *37*(10), 1033-1040.
26. Zhang, J., Chen, S., Yang, R., & Yan, Y. (2010). Biodiesel production from vegetable oil using heterogenous acid and alkali catalyst. *Fuel*, *89*(10), 2939-2944.
27. Zhang, X., Shanmugam, K. T., & Ingram, L. O. (2010). Fermentation of glycerol to succinate by metabolically engineered strains of Escherichia coli. *Applied and environmental microbiology*, *76*(8), 2397-2401.
28. Zheng, P., Zhang, K., Yan, Q., Xu, Y., & Sun, Z. (2013). Enhanced succinic acid production by Actinobacillus succinogenes after genome shuffling. *Journal of Industrial Microbiology and Biotechnology*, *40*(8), 831-840.
29. Saito, T., & Noda, H. (2017). "Challenges in succinic acid production from waste glycerol and other renewable carbon sources." Bioresource Technology, 228, 302-309
30. Lu, J., Lv, Y., Jiang, Y., Wu, M., Xu, B., Zhang, W.,& Jiang, M. (2020). Consolidated bioprocessing of hemicellulose-enriched lignocellulose to succinic acid through a microbial cocultivation system. *ACS Sustainable Chemistry & Engineering*, *8*(24), 9035-9045.
31. Dai, Z., Guo, F., Zhang, S., Zhang, W., Yang, Q., Dong, W., & Xin, F. (2020). Bio-based succinic acid: an overview of strain development, substrate utilization, and downstream purification. *Biofuels, Bioproducts and Biorefining*, *14*(5), 965-985.

32. Zhang, W., Zhang, T., Song, M., Dai, Z., Zhang, S., Xin, F., & Jiang, M. (2018). Metabolic engineering of Escherichia coli for high yield production of succinic acid driven by methanol. *ACS Synthetic Biology*, *7*(12), 2803-2811.
33. Song, H., & Lee, S. Y. (2006). Production of succinic acid by bacterial fermentation. *Enzyme and microbial technology*, *39*(3), 352-361.
34. Cheng, K. K., Zhao, X. B., Zeng, J., Zhang, J. A. (2012). Biotechnological production of succinic acid: Current state and perspectives. Biofuels, Bioproducts, and Biorefining, 6(3), 302–318.
35. D'ambrosio, S., Alfano, A., & Cimini, D. (2021). Production of succinic acid from Basfia succiniciproducens. *Frontiers in Chemical Engineering*, *3*, 785691.

Review Paper

VALORIZATION OF WASTE GLYCEROL FOR DIHYDROXYACETONE SYNTHESIS: PATHWAYS AND SUSTAINABILITY

Dr. S. Chozhavendhan, B Vijayambika[*], M Kiruthika

Vivekanandha College Of Engineering For Women(Autonomous)
**Corresponding Author: B Vijayambika*
Email id:vijayambika1111@gmail.com

ABSTRACT

Valorization of waste glycerol for dehydroxyacetone (DHA) production is a potential method for producing a valuable molecule with a wide range of applications in the pharmaceutical and cosmetic sectors. *Gluconobacter oxydans* or *Gluconobacter frateurii, Gluconacetobacter xylinus* are frequently involved in the microbial synthesis of DHA. These microorganisms have a high selectivity for converting glycerol to DHA, which uses crude glycerol as a carbon source. It explores the methods and processes for converting waste glycerol into DHA. It demonstrates a method that uses microbial bioconversion, followed by biotransformation, cell immobilization, and seed culture. The utilization of waste glycerol in DHA manufacturing provides a more sustainable and ecologically friendly alternative to standard processes. The analysis emphasizes the potential of this strategy to minimize waste while producing a valuable compound, thus contributing to a more circular economy. The synthesis and sustainability of DHA derived from waste glycerol are thoroughly reviewed.

KEYWORDS: *Microbial fermentation, Crude glycerol, Gluconobacter oxydans, Gluconobacter frateurii, Gluconacetobacter xylinus, sustainable production*

1. INTRODUCTION

Dihydroxyacetone is simply known as glycerine or ketotriose with a molecular formula $C_3H_6O_3$ [4]. Dihydroxyacetone (DHA) holds a significant place in market due to its application in the pharmaceutical and cosmetic industries[34]. Additionally, it is utilized as a raw material to make the polymers like 1,2-propylene glycol, 1,3-propanediol, lactic acid, methotrexate, surfactants, and ingredients in sun tanning products. Crude glycerol generated as byproducts during biodiesel production[43]. serves as an excellent carbon[52] and energy source for the microbial production for the DHA[67] .However, DHA can also be produced through chemical and catalytic oxidation fermentation over Gluconobacter oxydans or by biotransformation of glycerol[9].Several microbial species have been discovered as potential DHA producers from different substrates. Gluconobacter strains are the most commonly employed microorganisms (in both laboratory and industrial settings) for producing DHA from glycerol[15].*Gluconobacter oxydans* cells have been used commercially for many decades[4]. In addition to glycerol, it can employ additional carbon substrates such as d-sorbitol, d-fructose, and d-glucose[22]. Oxidative metabolism produces compounds such as Vitamin C, d-gluconic acid, ketogluconic acid, xylitol, vinegar, and so on[14].

2. DIHYDROXYACETONE

Dihydroxyacetone (DHA, 1,3-dihydroxy-2-propanone, C3H6O3) is a well-known ketotriose that is used as a raw material to make several important industrial chemicals, including lactic acid, 1,2-propylene glycol, 1,3-propanediol, methotrexate, surfactants, and others[32]. Gluconobacter bacteria are the only way to manufacture DHA on an industrial scale through microbial conversion from glycerol[23]. The usage of crude glycerol, an industrial by-product, has been investigated as a mean for its value and contribution to circularity, despite the fact that DHA is often made using pure glycerol. It is possible to transform the waste glycerol produced during the synthesis of biodiesel into a valuable product such as dihydroxyacetone (DHA). In the cosmetics sector, DHA is frequently used as a colorant in sunless tanning products. Given the rising sales of sunless tanning products, there is a market need for DHA. Primary active component of sunless tanning skin care products and acrucial building block f or the production of numerous excellent chemicals, 1,3Dihydroxyacetone (DHA), is made ind ustrialy by microbially fermenting glycerol over Gluconobacter oxydans. Figure 1 illustrate

Application of Dihydroxyacetone

Microbial fermentation methods have certain disadvantages, including limited productivity & high production costs, even though they can offer great selectivity to DHA.

Furthermore, the growth of bacteria is inhibited by both glycerol and DHA. There are several applications for dihydroxyacetone (DHA) in the food[7], pharmaceutical[51], medicine[26],and cosmetics[42] industries. The overproduction of glycerol by the biodiesel sector makes its production from glycerol financially intriguing. Dihydroxyacetone is produced via two recognized processes: chemical and microbiological[25]. Glycerol can be catalytically oxidized or condensed with calcium carbonate to initiate the chemical synthesis[19]. Microorganisms with high glycerol dehydrogenase activity, particularly acetic acid bacteria, can oxidize glycerol microbiologically as an alternative to chemically synthesizing DHA[11]. DHA in a metabolically modified strain of Escherichia coli by means of heterologous expression of the gene for C. Glutamicum dihydroxyacetone. phosphate dephosphorylase utilizing glucose as the only carbon source To the best of our knowledge, the presence of this gene in any other organism has not been found, and it was recently detected in C. glutamicum . The primary precursor is dihydroxyacetone phosphate (DHAP), which undergoes a single-step biochemical process with cghdpA to dephosphorylate to DHA[56].

3.CRUDE GLYCEROL AS RAW MATERIAL

Fossil fuels continue to be the basic source of energy source and chemical production globally. The conveyance sector now consumes 20% of global energies and is anticipated to expand due to population expansion and economic growth. Currently, it accounts for 25% energy-related emissions of carbon dioxide, with crude oil accounting for 95%. That depletion of fossil fuels, market energies monopolies, environmental degradation, and social concerns have led to increased interest in sources of renewable energy among scientists and the market.Government measures, such as the European Union's aim of 20% renewable energy consumption by 2020 and 10 % liquid biofuels in transportation, have contributed to progress in this area. Ethanol and biodiesel are the two most important biofuels. Biodiesel is regarded as a benign, renewable, and biodegradable resource[48].Biodiesel is a sustainable, biodegradable, non-toxic alternative to diesel fuel with comparable qualities. Since 2006, biodiesel has expanded beyond its traditional use in road transport to include aviation, energy, cooking, and maritime transport.Approximately 10% of biodiesel production is crude glycerol, which comprises impurities such as water, inorganic salts, methanol, fatty acids, esters, and more[58]. The glycerol (GLY) percentage ranges from 40% to 88 wt%. Varies according to the oil type and method used. Crude glycerol's purity level falls below technical (88 wt% to 98 wt%) and pharmaceutical (> 98 wt%) standards, making it a byproduct with little economic use[48] approximately 1 kg of crude glycerol is produced for 9 kg of biodiesel[32].Directly adding glycerol to fuels can create engine difficulties due to breakdown and polymerization processes. Table 1 Illustrate Common Parameters of Crude Glycerol. Crude glycerol is unstable for use in a variety of applications, including cosmetics, food, and pharmaceutical products[35].Previously, crude glycerol was processed as a waste stream from the biodiesel industry and incinerated for energy recovery [53]. Increasing biodiesel production has resulted in an overabundance of crude glycerol and a low market price. Valorizing crude glycerol into added-value products is a goal for improving the financial and ecological competitiveness of the biodiesel sector, along with green process principles[58]. Biodiesel manufacturing can be cost-effective by recycling methanol or ethanol from distillation processes. Organic matter can be recycled as a source material for acidic esters. Ash is typically made up of polluted sodium, calcium, or potassium sulfate, with proven market value. Separating contaminants may be feasible for large-scale biodiesel plants with sufficient capital.

4.PATHWAYS FOR GLYCEROL CONVERSION TO DIHYDROXYACETONE

4.1. Fermentation - *Gluconobacter oxydans*

Microorganism and Media preparation

G. oxydans ZJB09113, a mutant strain of G. oxydans ZJB-605, was used to produce DHA.The GY medium had a pH of 6.0 and contained glucose, yeast extract, and agar per liter, whereas the MY medium had a pH of 4.9 and included yeast extract, glycerol, mannitol, K_2HPO_4, KH_2PO_4, $MgSO4{\bullet}7H_2O$, and $CaCO_3$.Each type of media was autoclaved[31].

Seed Culture Preparation

Transfer cells from slant cultures that have been incubated for 24 hours into 500 mL flasks filled with liquid media GY.A rotary shaker was then used to incubate the flasks.Using sterile liquid media MY, 500 mL flasks were contaminated with seed culture. A rotary shaker was then used to incubate the flasks. Cells were extracted using centrifugation[14].

Cell Immobilization

Round porous ceramics were utilized as carriers for cell adsorption. The porous ceramics had a diameter of 10mm, a specific weight of 2.4 g/cm^3, porosity ranging from 30% to 40%, and pore diameters of 30, 40, and 50 µm. Before usage, the porous ceramics were cleaned with H_2O_2 by gently agitating at a specific temperature for 30 minutes. After decanting the liquid, the beads were rinsed with distilled water and dried overnight. The sterile liquid medium MY was introduced to the aseptic bubble column bioreactor, which was packed with porous ceramics, and 50 mL of seed culture was injected. The broth was withdrawn after 24 hours of aeration-based cell culture[30].

Biotransformation process

The aeration rate was measured and manually controlled with a flow meter equipped with a needle valve. Water circulation in the column's outer jacket controlled the temperature during operation. Culture pH was manually adjusted by applying NaOH/HCl. The foam was controlled by using polyethylene glycol as an antifoam agent. During the trials, the bioreactor was filled with an appropriate volume of sterile water [14]. Use a continuous glycerol and urea feeding method to maintain glycerol concentration. To examine the impact of aeration rate on DHA synthesis in the bubble column bioreactor, various aeration rates were chosen while other variables remained constant[33]. To keep the glycerol concentration stable, use a continuous feeding strategy that includes urea.Check the glycerol and DHA levels.To avoid product inhibition, stop the biotransformation when the DHA concentration reaches around 90-100 g/L. Remove the DHA solution and restart the biotransformation process [14].

Preparation of Free Extract

The wet *G. oxydans* biomass was suspended in sterile distilled water and ultrasonically disintegrated at 4°C using an Omni Ruptor 4000 (OMNI International, Kennesaw, GA, USA). The cell extract was centrifuged to eliminate debris, and the supernatant, containing a variety of cell enzymes, was saved for future use[6].

Free and immobilized cell extract biotransformation

The method for the cell extracts was similar to that for the whole cells, but the pH had been adjusted to 7.5[14] .

4.2. Oxidation

Initial step in the purification of crude glycerol is the elimination of ash and methanol. Fig.2 illustratePathways for Glycerol Conversion to Dihydroxyacetone. Oxidation of glycerol using a technological iron zeolite catalyst, which was synthesized through pelletization in a continuous gas-phase plug-flow reactor.Condensation under ambient conditions to separate the DHA/water-containing stream from the other reaction gasses. Pure DHA crystals are recovered when five-effect evaporative crystallization of DHA ends under vacuum (simply referred to as crystallization). Cleaning as well as drying the crystals of DHA[34].

4.3.Fermentation - *Gluconobacter frateurii*

Gluconobacter frateurii is used in the microbial synthesis of dihydroxyacetone (DHA), which comprises many essential stages. First, a specific strain of bacteria, *Gluconobacter frateurii* CGMCC 5397, is grown on biodiesel-derived crude glycerol as its primary carbon source. The identified *Gluconobacter frateurii* efficiently produces dihydroxyacetone from biodiesel-derived crude glycerol[65]. A fermentation medium comprising crude glycerol, corn steep liquor (CSL) as a source of nitrogen, and CaCO3 is inoculated with a 5% seed culture of this bacteria once the pH has been adjusted to 6.0[41]. In shake flasks, the fermentation process is carried out at 30°C with agitation at 200–220 rpm[65]. The oxygen transfer coefficient , a crucial factor in the generation of DHA, is controlled in bioreactors by adjusting airflow and agitation[61]. After 48 hours, batch fermentation usually produces about 80 g/L of DHA with an initial glycerol concentration of about 100 g/L. Feed-batch fermentation uses a lower beginning glycerol concentration of 20-60 g/L, and sterilized crude glycerol is fed throughout the process to maintain glycerol concentrations between 5-25 g/L.DHA concentrations of up to 175 g/L can be obtained using fed-batch fermentation and regulation of kLa[65]. Higher kLa values are often associated with improved glycerol utilization and enhanced DHA synthesis. Baffled flasks have a higher kLa than non-baffled flasks. Furthermore, CSL is utilized as a low-cost nitrogen supply, and it is critical to control salt chloride levels, which can hinder cell [61]. The fed-batch approach is employed to circumvent the inhibitory effects of high starting glycerol concentrations [65].

4.4. Fermentation - *Gluconacetobacter xylinus*

To activate glycerol dehydrogenase (GlyDH), *Gluconacetobacter xylinus* bacteria are first cultivated at 28°C in a liquid medium containing glycerol, yeast extract, and ammonium sulfate until OD600 ≈ 1.2. After centrifuging, cleaning, and suspending the bacterial biomass in sterile water, protease inhibitors and lysozyme are added to break down the cell walls enzymatically for two hours at 37°C [36]. Centrifugation is used to separate the resultant cell suspension into precipitate and supernatant, or it can be employed immediately. After combining these cell fractions with a sodium alginate solution, the mixture is dripped into a calcium chloride solution to create beads, which are then rinsed and allowed to solidify for 20 minutes at 4°C. The immobilized material is moved to a glycerol-containing production medium for biotransformation, and the pH is set to 5.0 for immobilized G. xylinus cells or 7.5 for cell preparations[55]. The biotransformation process is carried out using aeration at 200 cycles per minute at 28°C for immobilized cells or 23°C for cell preparations. For best DHA generation, the initial glycerol content should be 30 g/l. The procedure can be run for 4 or 20 days. Glycerol concentration is assessed refractometrically, while DHA concentration is ascertained colorimetrically with 3,5-dinitrosalicylic acid. Lastly, using the measured concentrations of glycerol and DHA, the biotransformation yield is computed. Catalytic activity is higher in the cell precipitate fraction[36].

5. MICROORGANISMS USED IN DHA PRODUCTION

The gram-negative bacterium *Gluconobacter oxydans* is a member of the *Acetobacteraceae* family. As an obligatory aerobe, *G. oxydans* uses oxygen as the terminal electron acceptor in its respiratory metabolism. The genus *Gluconobacter* is crucial to industry because it produces dihydroxyacetone from glycerol, D-gluconic acid, 5-keto-, and 2-ketogluconic acids from D-glucose, and L-sorbose from D-sorbitol.*Gluconobacter oxydans* ATCC 621 has been used in the kinetic investigations and optimization for the synthesis of DHA from glycerol. Ohrem and Voss (1996) investigated how glycerol affected *Gluconobacter oxydans* growth and DHA synthesis. They demonstrated that elevated glycerol concentrations had no effect on DHA metabolism or glycerol oxidation[8]. Table 2 illustrates the characteristic of *Gluconobacter Oxudense.*Even though *Gluconobacter* is already employed in industry to convert glycerol to DHA, there are still certain obstacles to be removed. The duration of DHA exposure causes an exponential decline in *Gluconobacter*viability , and the rate of decline is influenced by DHA concentration[45]. Additionally, both the specific growth rate of A high starting glycerol level may affect *G. oxydans* and the particular rate of DHA generation. Nevertheless, compared to glycerol DHA exhibits greater influence in producing an inhibitory effect on microbial development. Due to the robustness of glycerol dehydrogenase activity, the formation of this product persists for a while despite cell damage brought on by DHA accumulation in the media[5].*Gluconacetobacter xylinus* is an acetic acid bacteria that participates in the biotransformation of glycerol into dihydroxyacetone. The bacteria is stored in a growth medium containing Wort Agar, acetic acid, and ethyl alcohol at a pH of 5.0, cultured at 28°C for 24-48 hours, and replenished every 21 days. *G. xylinus* can be utilized to generate dihydroxyacetone (DHA) from glycerol via immobilized cell preparations. Glycerol dehydrogenase is used to biotransform glycerol to dihydroxyacetone in *G. xylinus* cells[36]. *Gluconobacterfrateurii* is a bacterium known for its capacity to manufacture dihydroxyacetone (DHA) from glycerol. *G. frateurii* CGMCC 5397, a recently identified strain, can produce DHA through shake-flask fermentation of biodiesel-derived crude glycerol. *G. frateurii* is closely related to other *Gluconobacter* species, including *G. oxydans, G. thailandicus, G. cerinus*, and *G. japonicus*, according to phylogenetic study[65].

6. SUSTAINABILITY AND ENVIRONMENTAL IMPACT OF DHA FROM CRUDE GLYCEROL

DHA can be produced from glycerol using the acetic acid bacterium *Gluconobacter oxydans*, which is both cost effective and ecologically friendly [38]. Raw glycerol is a byproduct of the biodiesel and oleochemical industries[12]. Raw glycerol is refined and used as a feedstock for several industrial compounds. DHA can be synthesized from untreated raw glycerol containing methanol and alkali metals[50]. Nevertheless, the purification process is costly and energy-intensive. The resultant quantity and purity of DHA should be improved further.The efforts to enhance industrial DHA production primarily involve the search for new methods to prevent or reduce inhibitory effects of the substrate/product[29,14], strain improvement in bioconversion[37,57], and optimization of process conditions[28,54]. The environmental benefits of using crude glycerol go beyond energy savings. If left unused, crude glycerol is usually processed in a biological waste water treatment facility or burnt in a waste-to-energy plant[39]. Life cycle analysis (LCA) has revealed that chemocatalytic DHA generation from crude

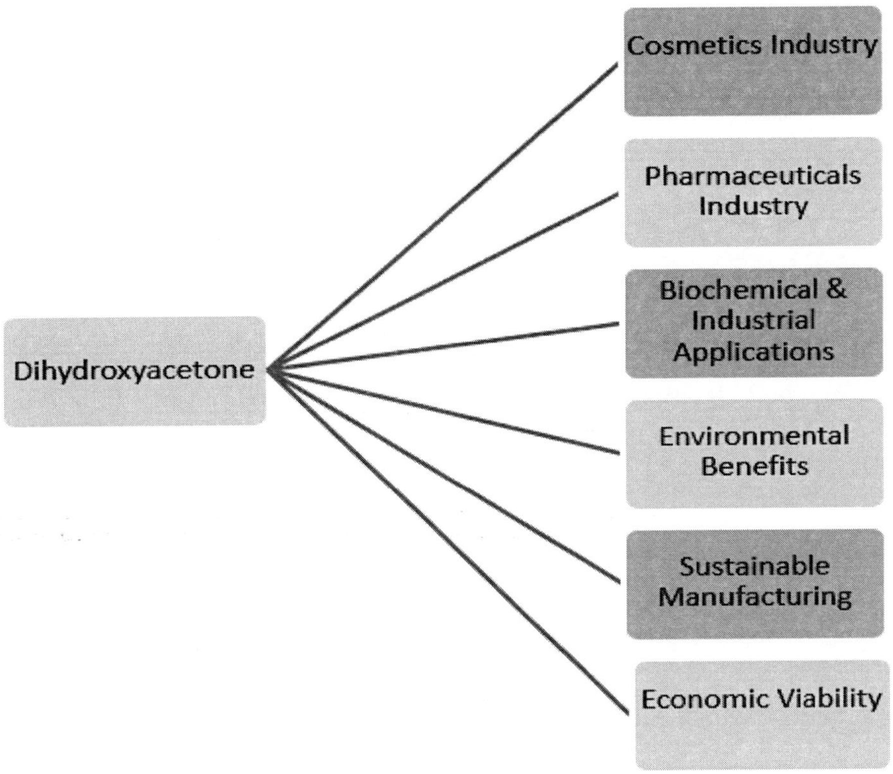

Figure 1. Application of Dihydroxyacetone

Table 1. Common Parameters of Crude Glycerol[3,13]

Parameters	Crude Glycerol
Glycerol Content	23 – 63%
Methanol	6 – 13 %
Water	1 – 29 %
Ash	2 – 6 %
Free Fatty acid	1 – 3 %
Ph	6.9
Color	Dark Brown

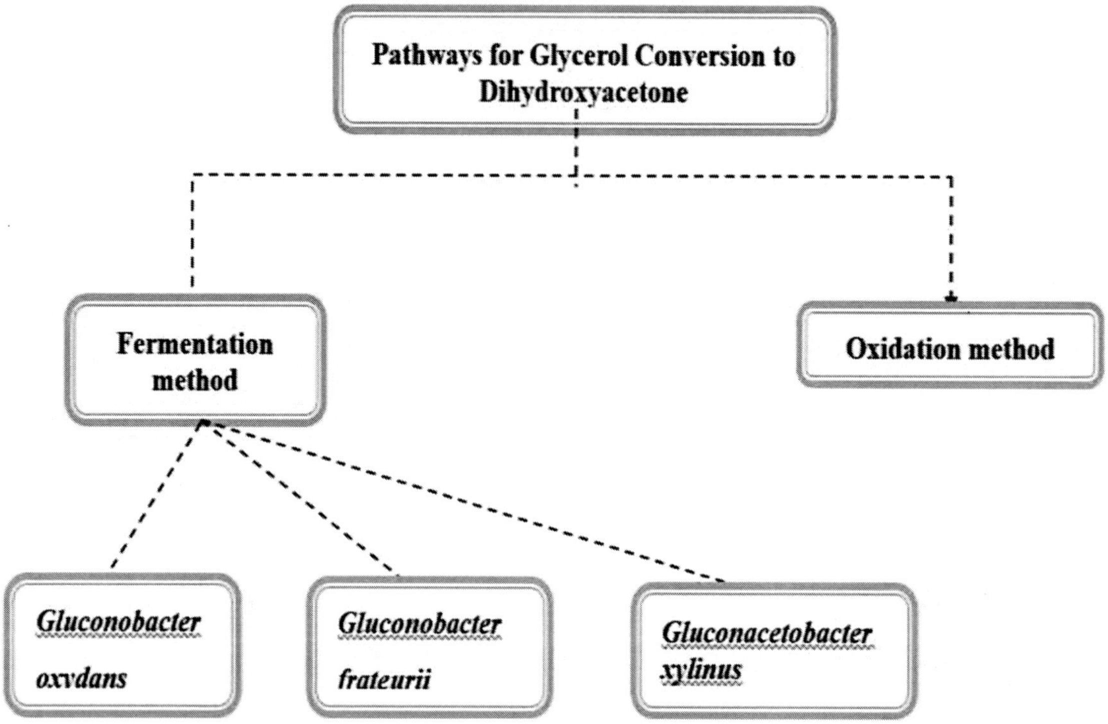

Figure 2.Pathways for Glycerol Conversion to Dihydroxyacetone

glycerol has a significantly lower environmental impact than traditional biocatalytic approaches[2]. This difference is due to the greater DHA concentrations achieved during the chemocatalytic process, which allows for the use of less energy-intensive separation methods.Furthermore, the chemocatalytic process runs constantly, which is more efficient than batch operations [17]. The chemocatalytic approach also has economic advantages, such as lower energy usage and the potential to integrate heat recovery systems[1]. Immobilized G. oxydans cells are used in a specific reactor system to boost production rates [40]. Optimizing process parameters like pH and aeration to increase conversion rates[60]. Genetically altering strains to improve DHA tolerance and output[64].Investigating alternate purifying technologies to produce a purer product while lowering downstream processing expenses[62]. Ultimately, the viability of DHA production from crude glycerol, regardless of method, is dependent on a commitment to constant development and optimization. More study into both biocatalytic and chemocatalytic techniques is required to determine the most effective and sustainable strategies for increasing yield, reducing waste, and lowering overall environmental impact.

Table 2. Characteristics of *Gluconobacter Oxudense, Gluconobacter Frateurii, Gluconobacter Xylinus*

Characteristics	*Gluconobacter Oxudense*	*Gluconobacter Frateurii*	*Gluconobacter Xylinus*
Scientific Classification	Acetic acid bacteria(AAB)	Acetic acid bacteria(AAB)	Acetic acid bacteria(AAB)
Morphology	Rod shaped ,single cells	Rod shaped , single cells or pairs	Rod shaped ,single cells or slender
Color	White to off white colonies	Off white or pale	Off white to pale cream
Cell Arrangement	Single cells or in pairs	Single cells or in pairs	Typically found in clusters or chains
Gram Stain	Gram negative	Gram negative	Gram negative
Oxygen Requirement	Aerobic	Aerobic	Aerobic
Fermentation Product	Dehydroxyacetone(DHA), Gluconic acid	Dehydroxyacetone(DHA)	Cellulose acitic acid
Metabolism	Oxidizies glucose to gluconic acid, then DHA	Oxidizies glucose to gluconic acid, then DHA	Acetic acid production ,cellulose synthesis
Appearance on Agar	White, smooth colonies	White, smooth colonies	White, Fibrous,or slimy colonies
Industrial use	DHA production in biotechnology	DHA production in biotechnology	Cellulose production , potential for DHA production

7. CONCLUSION

It is theoretically and financially possible to produce dihydroxyacetone (DHA) from waste glycerol, especially when utilizing microbial fermentation with Gluconobacter oxydans. This process provides a viable way to value a substantial byproduct of the manufacturing of biodiesel. The possible to manufacture significant amounts of DHA from easily accessible crude glycerol while maintaining profitability even in the face of shifting DHA market prices. DHA yield is greatly increased by process improvement, such as reusing the n-butanol stream. Although handling chemicals throughout the process entails some modest safety concerns, these may be controlled with the right safety precautions. Iron zeolites are used in alternative chemocatalytic processes that exhibit promise and have less of an adverse environmental impact than enzymatic techniques. Although the use of immobilized cells or cell extracts can overcome these problems and produce even better DHA yields, more research is required to fully understand the impact of initial glycerol concentration because higher concentrations can cause substrate inhibition. Indirect oxidation method has the ability to produce DHA on a big scale and gives good yields. The overall conclusion is that there are several practical ways to transform waste glycerol into the valuable substance DHA.

REFERENCES

1. Ribera, A., Arends, I. W. C. E., De Vries, S., Pérez-Ramırez, J., & Sheldon, R. A. (2000). Preparation, characterization, and performance of FeZSM-5 for the selective oxidation of benzene to phenol with N2O. *Journal of Catalysis*, *195*(2), 287-297.
2. Zelikin, A. N., Zawaneh, P. N., & Putnam, D. (2006). A functionalizable biomaterial based on dihydroxyacetone, an intermediate of glucose metabolism. *Biomacromolecules*, *7*(11), 3239-3244.
3. Abdul Raman, A. A., Tan, H. W., & Buthiyappan, A. (2019). Two-step purification of glycerol as a value added by product from the biodiesel production process. *Frontiers in chemistry*, *7*, 774.
4. da Silva, G. A. R., Oliveira, S. S. D. S., Lima, S. F., do Nascimento, R. P., Baptista, A. R. D. S., & Fiaux, S. B. (2022). The industrial versatility of Gluconobacter oxydans: current applications and future perspectives. *World Journal of Microbiology and Biotechnology*, *38*(8), 134.
5. Bauer, R., Katsikis, N., Varga, S., & Hekmat, D. (2005). Study of the inhibitory effect of the product dihydroxyacetone on Gluconobacter oxydans in a semi-continuous two-stage repeated-fed-batch process. *Bioprocess and biosystems engineering*, *28*, 37-43.
6. Stasiak-Różańska, L., Berthold-Pluta, A., & Dikshit, P. K. (2018). Valorization of waste glycerol to dihydroxyacetone with biocatalysts obtained from Gluconobacter oxydans. *Applied Sciences*, *8*(12), 2517.
7. Bicker, M., Endres, S., Ott, L., & Vogel, H. (2005). Catalytical conversion of carbohydrates in subcritical water: A new chemical process for lactic acid production. *Journal of Molecular Catalysis A: Chemical*, *239*(1-2), 151-157.
8. Bories, A., Claret, C., & Soucaille, P. (1991). Kinetic study and optimisation of the production of dihydroxyacetone from glycerol using Gluconobacter oxydans. *Process Biochemistry*, *26*(4), 243-248.
9. Choi, P. G., Nunotani, N., & Imanaka, N. (2018). Efficient production of dihydroxyacetone from glycerol over a Pt/CeO2-ZrO2-Bi2O3/SBA-16 catalyst. *Journal of Asian Ceramic Societies*, *6*(4), 368-373.
10. Choquenet B, Couteau C, Paparis E, Coiffard LJM (2009) J Dermatol 36:587–5918.

11. Claret, C., Bories, A., & Soucaille, P. (1992). Glycerol inhibition of growth and dihydroxyacetone production by Gluconobacter oxydans. *Current Microbiology, 25*, 149-155.
12. Da Silva, G. P., Mack, M., & Contiero, J. (2009). Glycerol: a promising and abundant carbon source for industrial microbiology. *Biotechnology advances, 27*(1), 30-39.
13. Sun, D., Yamada, Y., Sato, S., & Ueda, W. (2017). Glycerol as a potential renewable raw material for acrylic acid production. *Green Chemistry, 19*(14), 3186-3213.
14. Dikshit, P. K., & Moholkar, V. S. (2016). Kinetic analysis of dihydroxyacetone production from crude glycerol by immobilized cells of Gluconobacter oxydans MTCC 904. *Bioresource Technology, 216*, 948-957.
15. Dikshit, P. K., & Moholkar, V. S. (2016). Kinetic analysis of dihydroxyacetone production from crude glycerol by immobilized cells of Gluconobacter oxydans MTCC 904. *Bioresource Technology, 216*, 948-957.
16. Enders, D., Voith, M., & Lenzen, A. (2005). The dihydroxyacetone unit—A versatile C3 building block in organic synthesis. *Angewandte Chemie International Edition, 44*(9), 1304-1325.
17. Navalon, S., de Miguel, M., Martin, R., Alvaro, M., & Garcia, H. (2011). Enhancement of the catalytic activity of supported gold nanoparticles for the Fenton reaction by light. *Journal of the American Chemical Society, 133*(7), 2218-2226.
18. Ga¨tgens C, Degner U, Bringer-Meyer S, Herrmann U (2007) Appl Microbiol Biotechnol 76:553–559
19. Gehrer E, Harder W, Vogel H, Knuth B, Ebel K, Groening C (1995) USA Patent nr 5410089
20. Lari, G. M., Mondelli, C., & Pérez-Ramírez, J. (2015). Gas-phase oxidation of glycerol to dihydroxyacetone over tailored iron zeolites. *ACS Catalysis, 5*(3), 1453-1461.
21. Gupta, A., Singh, V. K., Qazi, G. N., & Kumar, A. (2001). Gluconobacter oxydans: its biotechnological applications. *Journal of molecular microbiology and biotechnology, 3*(3), 445-456.
22. Gupta, A., Singh, V. K., Qazi, G. N., & Kumar, A. (2001). Gluconobacter oxydans: its biotechnological applications. *Journal of molecular microbiology and biotechnology, 3*(3), 445-456.
23. Gupta, A., Singh, V. K., Qazi, G. N., & Kumar, A. (2001). Gluconobacter oxydans: its biotechnological applications. *Journal of molecular microbiology and biotechnology, 3*(3), 445-456.
24. Habe, H., Shimada, Y., Yakushi, T., Hattori, H., Ano, Y., Fukuoka, T., ...& Sakaki, K. (2009). Microbial production of glyceric acid, an organic acid that can be mass produced from glycerol. *Applied and Environmental Microbiology, 75*(24), 7760-7766.
25. Hekmat, D., Bauer, R., & Fricke, J. (2003). Optimization of the microbial synthesis of dihydroxyacetone from glycerol with Gluconobacter oxydans. *Bioprocess and biosystems engineering, 26*, 109-116.
26. Henderson, P. W., Kadouch, D. J., Singh, S. P., Zawaneh, P. N., Weiser, J., Yazdi, S., ... & Spector, J. A. (2010). A rapidly resorbable hemostatic biomaterial based on dihydroxyacetone. *Journal of Biomedical Materials Research Part A: An Official Journal of The Society for Biomaterials, The Japanese Society for Biomaterials, and The Australian Society for Biomaterials and the Korean Society for Biomaterials, 93*(2), 776-782.
27. Herrmann, U., & Sahm, H. (2005). Application of Gluconobacter oxydans for biotechnologically relevant reactions.
28. Hu, Z. C., Tian, S. Y., Ruan, L. J., & Zheng, Y. G. (2017). Repeated biotransformation of glycerol to 1, 3-dihydroxyacetone by immobilized cells of Gluconobacter oxydans with glycerol-and urea-feeding strategy in a bubble column bioreactor. *Bioresource technology, 233*, 144-149.
29. Hu, Z. C., & Zheng, Y. G. (2011). Enhancement of 1, 3-dihydroxyacetone production by a UV-induced mutant of Gluconobacter oxydans with DO control strategy. *Applied biochemistry and biotechnology, 165*, 1152-1160.
30. Hu, Z. C., Tian, S. Y., Ruan, L. J., & Zheng, Y. G. (2017). Repeated biotransformation of glycerol to 1, 3-dihydroxyacetone by immobilized cells of Gluconobacter oxydans with glycerol-and urea-feeding strategy in a bubble column bioreactor. *Bioresource technology, 233*, 144-149.
31. Jain, V. K., Tear, C. J. Y., & Lim, C. Y. (2016). Dihydroxyacetone production in an engineered Escherichia coli through expression of Corynebacterium glutamicum dihydroxyacetone phosphate dephosphorylase. *Enzyme and Microbial Technology, 86*, 39-44.
32. Khanna, S., Goyal, A., & Moholkar, V. S. (2012). Microbial conversion of glycerol: present status and future prospects. *Critical reviews in biotechnology, 32*(3), 235-262.
33. Dikshit, P. K., & Moholkar, V. S. (2019). Batch and repeated-batch fermentation for 1, 3-dihydroxyacetone production from waste glycerol using free, immobilized and resting Gluconobacter oxydans cells. *Waste and Biomass Valorization, 10*, 2455-2465.

34. Lari, G. M., Mondelli, C., Papadokonstantakis, S., Morales, M., Hungerbühler, K., & Pérez-Ramírez, J. (2016). Environmental and economic assessment of glycerol oxidation to dihydroxyacetone over technical iron zeolite catalysts. *Reaction Chemistry & Engineering*, *1*(1), 106-118.
35. Li, M. H., Wu, J., Liu, X., Lin, J. P., Wei, D. Z., & Chen, H. (2010). Enhanced production of dihydroxyacetone from glycerol by overexpression of glycerol dehydrogenase in an alcohol dehydrogenase-deficient mutant of Gluconobacter oxydans. *Bioresource technology*, *101*(21), 8294-8299.
36. Stasiak-Różańska, L., Błażejak, S., & Miklaszewska, A. (2011). Application of immobilized cell preparation obtained from biomass of Gluconacetobacter xylinus bacteria in biotransformation of glycerol to dihydroxyacetone. *Acta Scientiarum Polonorum Technologia Alimentaria*, *10*(1), 35-49.
37. Lin, X., Liu, S., Xie, G., Chen, J., Li, P., & Chen, J. (2016). Enhancement of 1, 3-dihydroxyacetone production from Gluconobacter oxydans by combined mutagenesis. *Journal of Microbiology and Biotechnology*, *26*(11), 1908-1917.
38. Lin, X., Liu, S., Xie, G., Chen, J., Li, P., & Chen, J. (2016). Enhancement of 1, 3-dihydroxyacetone production from Gluconobacter oxydans by combined mutagenesis. *Journal of Microbiology and Biotechnology*, *26*(11), 1908-1917.
39. Morales, M., Dapsens, P. Y., Giovinazzo, I., Witte, J., Mondelli, C., Papadokonstantakis, S., ...& Pérez-Ramírez, J. (2015). Environmental and economic assessment of lactic acid production from glycerol using cascade bio-and chemocatalysis. *Energy & Environmental Science*, *8*(2), 558-567.
40. Merfort, M., Herrmann, U., Ha, S. W., Elfari, M., Bringer-Meyer, S., Görisch, H., & Sahm, H. (2006). Modification of the membrane-bound glucose oxidation system in Gluconobacter oxydans significantly increases gluconate and 5-keto-d-gluconic acid accumulation. *Biotechnology Journal: Healthcare Nutrition Technology*, *1*(5), 556-563.
41. Metsoviti, M., Paraskevaidi, K., Koutinas, A., Zeng, A.P., Papanikolaou, S., Production of 1,3-propanediol, 2,3-butanediol and ethanol by a newly isolated Klebsiella oxytoca strain growing on biodiesel-derived glycerol based media. Process Biochem 2012.. 47, 1872-1882
42. Mishra, R., Jain, S. R., & Kumar, A. (2008). RETRACTED: Microbial production of dihydroxyacetone. *Biotechnology Advances*, *26*(4), 293-303.
43. Tiong, A. N. T., Cheonh, P. Y. Y., Chua, W. F., Kolloju, N., Ting, J. M., & Thian, J. J. (2024). Valorization of waste glycerol into value-added dihydroxyacetone via microbial fermentation: feasibility study. *Chemical Papers*, 1-16.
44. Obeid, O. A., Jamal, Z. M., Hwalla, N., & Emery, P. W. (2006). The effect of glutamine and dihydroxyacetone supplementation on food intake, weight gain, and postprandial glycogen synthesis in female Zucker rats. *Nutrition*, *22*(7-8), 794-801.
45. Ohrem, H. L., & Voβ, H. (1995). Inhibitory effects of dihydroxyacetone on Gluconobacter cultures. *Biotechnology letters*, *17*, 981-984.
46. Painter, R. M., Pearson, D. M., & Waymouth, R. M. (2010). Selective catalytic oxidation of glycerol to dihydroxyacetone. *Angewandte Chemie (International ed. in English)*, *49*(49), 9456–9459.
47. Prust, C. (2004). Entschlüsselung des Genoms von Gluconobacter oxydans 621H-einem Bakterium von industriellem Interesse.
48. Quispe, C. A., Coronado, C. J., & Carvalho Jr, J. A. (2013). Glycerol: Production, consumption, prices, characterization and new trends in combustion. *Renewable and sustainable energy reviews*, *27*, 475-493.
49. Rogers, C. J. (2005). Spray-on tanning. *Aesthetic Surgery Journal*, *25*(4), 413-415.
50. Rywinska, A., Rymowicz, W., Zarowska, B., & Wojtatowicz, M. (2009). Biosynthesis of citric acid from glycerol by acetate mutants of Yarrowia lipolytica in fed-batch fermentation. *Food Technol Biotechnol*, *47*(1), 1-6.
51. Schmid D, Belser E, Zulli F (2007) Cosmet Toilet 6:55–6
52. Ahmad, M. S., Ab Rahim, M. H., Alqahtani, T. M., Witoon, T., Lim, J. W., & Cheng, C. K. (2021). A review on advances in green treatment of glycerol waste with a focus on electro-oxidation pathway. *Chemosphere*, *276*, 130128.
53. Skrzyńska, E., Wondołowska-Grabowska, A., Capron, M., & Dumeignil, F. (2014). Crude glycerol as a raw material for the liquid phase oxidation reaction. *Applied Catalysis A: General*, *482*, 245-257.
54. Sudarshan, B. L., & Sanjay, K. R. (2018). Optimization of fermentation conditions for production of 1, 3-dihydroxyacetone from glycerol obtained as a byproduct during biodiesel production. *Microbiol Res J Int*, *25*, 1-9.

55. Survase, S. A., Annapure, U. S., & Singhal, R. S. (2010). Gellan gum as immobilization matrix for production of cyclosporin A. *Journal of microbiology and biotechnology*, *20*(7), 1086-1091.
56. Jojima, T., Igari, T., Gunji, W., Suda, M., Inui, M., & Yukawa, H. (2012). Identification of a HAD superfamily phosphatase, HdpA, involved in 1, 3-dihydroxyacetone production during sugar catabolism in Corynebacterium glutamicum. *FEBS letters*, *586*(23), 4228-4232.
57. Tan, J., Yang, X., & Lu, W. (2019). Research of 1, 3-dihydroxyacetone production by overexpressing glycerol transporter and glycerol dehydrogenase. *Transactions of Tianjin University*, *25*, 549-558.
58. Trifoi, A. R., Agachi, P. Ş., & Pap, T. (2016). Glycerol acetals and ketals as possible diesel additives. A review of their synthesis protocols. *Renewable and Sustainable Energy Reviews*, *62*, 804-814.
59. Weiser, J. R., Zawaneh, P. N., & Putnam, D. (2011). Poly (carbonate-ester) s of dihydroxyacetone and lactic acid as potential biomaterials. *Biomacromolecules*, *12*(4), 977-986.
60. Wethmar, M., & Deckwer, W. D. (1999). Semisynthetic culture medium for growth and dihydroxyacetone production by Gluconobacter oxydans. *Biotechnology techniques*, *13*(4), 283-287.
61. Zheng, X. J., Jin, K. Q., Zhang, L., Wang, G., & Liu, Y. P. (2016). Effects of oxygen transfer coefficient on dihydroxyacetone production from crude glycerol. *brazilian journal of microbiology*, *47*, 129-135.
62. Xie G (2018) Method for preparing DHA dry powder through microbial fermentative liquid (2018)
63. Xu, X., Chen, X., Jin, M., Wu, X., & Wang, X. (2009). Advance in dihydroxyacetone production by microbial fermentation. *Sheng wu Gong Cheng xue bao= Chinese Journal of Biotechnology*, *25*(6), 903-908.
64. Xu X, He Y, Xia B, Zhao Q 1,3-dihydroxyacetone highyielding strain and construction method thereof (2014)
65. Liu, Y. P., Sun, Y., Tan, C., Li, H., Zheng, X. J., Jin, K. Q., & Wang, G. (2013). Efficient production of dihydroxyacetone from biodiesel-derived crude glycerol by newly isolated Gluconobacter frateurii. *Bioresource technology*, *142*, 384-389.
66. Zhang, L., Yang, Y., Sun, J. A., Shen, Y., Wei, D., Zhu, J., & Chu, J. (2010). Microbial production of 2, 3-butanediol by a mutagenized strain of Serratia marcescens H30. *Bioresource Technology*, *101*(6), 1961-1967.
67. Zheng, Z., Luo, M., Yu, J., Wang, J., & Ji, J. (2012). Novel process for 1, 3-dihydroxyacetone production from glycerol. 1. Technological feasibility study and process design. *Industrial & engineering chemistry research*, *51*(9), 3715-3721.

Research Paper

TRANSFORMING WHOLE BROKEN EGG IN TO BIOFERTILIZER

Dr. Rajmehala M[1], Devika R[2], Dhinishiya M[2], SharmilaR[2], Swathi S[2]

[1]Associate professor, Department of Biotechnology, Vivekananda College of Engineering for Women (Autonomous), Tiruchengode-637205 Namakkal DT, Tamilnadu India
[2]Department of Biotechnology, Vivekananda College of Engineering for Women (Autonomous), Tiruchengode-637205, Namakkal DT, Tamilnadu India
**Corresponding Address: DrRajmehalaM*
Email id:rajamehala@vcew.ac.in

ABSTRACT

Our idea focuses on producing biofertilizer from broken eggs. By utilizing these broken eggs, we transform them into an effective and nutrient-rich biofertilizer. This approach offers an efficient solution to environmental pollution and provides benefits to both farmers and poultry owners. It aims to reduce egg wastage by converting waste into wealth. Namakkal, being the second-largest poultry producer in India, generates around 5-6 crore eggs daily. Approximately 1% of these eggs, nearly 5,00,000, are broken during collection and transportation. Instead of discarding them as waste, we convert them into a valuable biofertilizer. The process of turning cracked eggs into biofertilizer is an environmental friendly method to control pollution and manage waste. This biofertilizer is rich in calcium, potassium, phosphorus, and other organic compounds, enhancing agricultural production and soil health. The steps involved in producing biofertilizer from broken eggs include collecting the eggs from poultry farms, frying and drying them to remove moisture, and then crushing them into powder form. Neem cake is added to the egg powder, acting as a biopesticide to protect crops from insects and pests, reducing the need for chemical pesticides. This not only promotes healthier crops but also minimizes the environmental impact of chemical use in agriculture. Economically, this approach provides a cost-effective alternative to chemical fertilizers. Farmers can save on the costs associated with purchasing commercial fertilizers, making this method financially advantageous. Furthermore, in regions like Namakkal, this initiative can boost the local economy by creating jobs and supporting both the agricultural and poultry sectors. It contributes to long-term soil health and helps maintain the ecological balance.

KEYWORDS: *Biofertilizer, Broken eggs, Environmental pollution, Farmers, Nutrient- rich, Calcium, Potassium, Phosphorus, Neem cake, Biopesticide.*

1. INTRODUCTION

In a groundbreaking effort to tackle environmental pollution and foster sustainable agricultural practices, the transformation of broken eggs into nutrient-rich biofertilizer presents a revolutionary solution. Namakkal, Tamil Nadu, renowned as India's second-largest poultry producer, faces a significant challenge with around 5,00,000 eggs breaking daily during collection and transportation. Rather than allowing these eggs to go to waste, this innovative initiative repurposes them into valuable biofertilizer, abundant in essential nutrients such as calcium, potassium, and phosphorus.

By converting waste into wealth, this method not only addresses the pressing issue of egg wastage but also brings a multitude of benefits to both farmers and poultry owners. The process of producing biofertilizer from broken eggs is an environmentally friendly approach that mitigates pollution while enhancing agricultural productivity and soil health. The steps involved are meticulous: collecting the broken eggs from poultry farms, frying and drying them to eliminate moisture, crushing them into a fine powder, and incorporating neem cake. Neem cake acts as a natural biopesticide, protecting crops from pests and reducing the dependency on chemical pesticides. This eco-conscious approach is not only beneficial for crop health but also minimizes the environmental impact of chemical use in agriculture. The biofertilizer, rich in organic compounds, supports robust plant growth and soil vitality, contributing to long-term agricultural sustainability. Economically, this method offers a cost-effective alternative to traditional chemical fertilizers, enabling farmers to save on commercial fertilizer costs and invest in more sustainable farming practices.

The poultry industry, one of the largest contributors to global food production, faces pressing challenges in sustainability, waste management, and nutrient utilization. Approximately 10-15% of eggs produced are discarded due to cosmetic imperfections, resulting in significant economic losses and environmental concerns. Whole chicken cracked eggs, rich in calcium, potassium, phosphorus, and organic compounds, present an untapped opportunity for value-added recycling. This study investigates the development and evaluation of a novel biofertilizer product incorporating whole chicken cracked eggs, aiming to reduce waste, enhance soil health, and promote sustainable agricultural practices. By exploring the technical, environmental, and economic feasibility of cracked egg- based biofertilizer, this research seeks to contribute to a more circular and resilient food system. Traditionally, the efficiency and profitability of poultry production rely heavily on high-quality feed formulations and effective waste management practices. However, the increasing costs of conventional fertilizers and environmental concerns surrounding their production have raised the need for more sustainable and cost-effective alternatives. An innovative solution to this challenge lies in the potential use of cracked chicken eggs as an alternative biofertilizer component. Cracked eggs, which are often discarded in egg processing facilities due to their damage, represent a significant waste stream despite containing high nutritional value. These eggs, typically unsuitable for human consumption, are rich in essential hat can benefit plant growth and soil health. With proper processing and incorporation, cracked eggs can offer a high-quality, nutrient-dense biofertilizer source capable of meeting the specific soil enrichment requirements of various crops. The process involves collecting the broken eggs, frying and drying them to remove moisture, crushing them into a fine powder, and mixing them with neem cake to act as a natural biopesticide. This biofertilizer supports robust plant growth, enhances soil vitality, and contributes to long-term agricultural sustainability. Economically, this method offers a cost-effective alternative to traditional chemical fertilizers, enabling farmers to save on commercial fertilizer costs and invest in more sustainable farming practices. In regions like Namakkal, this initiative has the potential to invigorate the local economy by creating job opportunities and bolstering the agricultural and poultry sectors. It represents a forward-thinking approach to waste management, turning a significant environmental challenge into a valuable resource. Ultimately, the transformation of broken eggs into biofertilizer exemplifies a harmonious blend of innovation, sustainability, and economic benefit, paving the way for a greener and more prosperous future in agriculture.

2. MATERIALS AND METHODS

Collection of sample

To collect the cracked eggs from poultry form in Namakkal DT, Tamil nadu.

Sample preparation

The sample preparation process commenced with the collection of whole chicken eggs, which were then weighed and thoroughly mixed to combine the shell, yolk, and white. The eggs were subsequently fried for 10 minutes over a gas stove flame, followed by a 5-minute cooling period at room temperature. Next, the eggs were dried in a hot air oven at 200°C for 25 minutes at medium heat, and then cooled for an additional 20 minutes at room temperature. Finally, the dried eggs were ground into a fine powder, weighed, and stored for further analysis. The process of egg powder preparation is summarized in flow chart as follow is shown in

Figure 1.

Process of Biofertilizer

The process of biofertilizer preparation is shown in figure 2 and figure 3.

Whole Egg Powder

Whole egg powder is a nutrient rich ingredient made from whole eggs that have been spary dried or freeze dried to remove the moisture. It is a rich source of protein, vitamins, and minerals making it an ideal ingredient for various applications. Whole egg powder contains approximately 45-50% protein, 35-40% fat, and 5% moisture. It is commonly used in poultry and animal feed, food processing, pharmaceuticals, and cosmetics. With a shelf life of up to 12 months, whole egg powder is a convenient and versatile ingredient.

3. MATERIALS AND METHODS

Collection of sample

To collect the cracked eggs from poultry form in Namakkal DT, Tamil nadu.

Sample preparation

The sample preparation process commenced with the collection of whole chicken eggs, which were then weighed and thoroughly mixed to combine the shell, yolk, and white. The eggs were subsequently fried for 10 minutes over a gas stove flame, followed by a 5-minute cooling period at room temperature. Next, the eggs were dried in a hot air oven at 200°C for 25 minutes at medium heat, and then cooled for an additional 20 minutes at room temperature. Finally, the dried eggs were ground into a fine powder, weighed, and stored for further analysis. The process of egg powder preparation is summarized in flow chart as follow is shown in

Figure 1.

Process of Biofertilizer

The process of biofertilizer preparation is shown in figure 2 and figure 3.

Whole Egg Powder

Whole egg powder is a nutrient rich ingredient made from whole eggs that have been spary dried or freeze dried to remove the moisture. It is a rich source of protein, vitamins, and minerals making it an ideal ingredient for various applications. Whole egg powder contains approximately 45-50% protein, 35-40% fat, and 5% moisture. It is commonly used in poultry and animal feed, food processing, pharmaceuticals, and cosmetics. With a shelf life of up to 12 months, whole egg powder is a convenient and versatile ingredient.

Crude Ash Determination

The IS 14827-2000 method determines crude ash in animal feed by incinerating a 5-10g sample at 500-600°C for 2-3 hours, removing organic matter and leaving behind inorganic ash. The resulting residue is weighed and calculated as a percentage of the original sample weight.

Crude Ash (%) = (Residue weight / Sample weight) x 100

Acid Insoluble Ash Determination

The IS 14826-2000 method determines acid-insoluble ash in animal feed by treating crude ash with hydrochloric acid, filtering, and weighing the residue. The process involves boiling crude ash with 10% HCl, filtering, washing, and re-igniting the residue at 500-600°C. The resulting weight represents acid-insoluble ash, calculated as a percentage of the original sample weight.

Acid-Insoluble Ash (%) = (Residue weight / Sample weight) x 100

Calcium determination

The AOAC 927.02 method determines calcium in animal feed using atomic absorption spectroscopy (AAS) or flame photometry. A 5-10g sample is ashed at 500-600°C, then digested with hydrochloric and nitric acids. Lanthanum chloride is added to reduce interference, and calcium is measured at 422.7 nm (AAS) or 554 nm (flame photometry).

Calcium (%) = (Concentration x Dilution factor) / Sample weight

Phosphorus determination

The AOAC 965.17 method determines phosphorus in animal feed through dry ashing, acid digestion, and colorimetric analysis. A 2-5g sample is ashed at 500-600°C, then digested with hydrochloric and nitric acids. The resulting phosphate is reacted with ammonium molybdate to form a blue complex, measured spectrophotometrically at 820 nm.

Phosphorus (%) = (Absorbance x Calibration factor) / Sample weight

Salt as chloride determination

The AOAC 943.01 method determines chloride (salt) in animal feed by water extraction and titration with silver nitrate solution. A 10-20g sample is extracted with water, filtered, and then titrated with AgNO3 until a persistent reddish-brown color indicates the formation of silver chloride.

Chloride (%) = (Titration volume x AgNO3 concentration x 0.3546) / Sample weight

Gross energy determination

Gross energy (GE) determination in animal feed is crucial for formulating balanced diets, evaluating feed quality, and predicting animal performance. GE represents the total energy content of the feed, expressed in units such as kcal/kg or MJ/kg. Direct methods like bomb calorimetry and oxygen bomb calorimetry measure heat released during combustion, while indirect methods like proximate analysis, near-infrared reflectance spectroscopy (NIRS), and empirical for mulas estimate GE based on feed composition and properties. Factors like moisture, crude protein, crude fat, fiber, and ash content affect GE, and regulatory standards from organizations like AOAC, ASTM, and ISO ensure accurate analysis. Understanding GE is essential for optimizing feed efficiency, ensuring animal health, and maintaining sustainable agricultural practices.

GE can be estimated using empirical formulas based on proximate analysis results:

GE (MJ/kg) = 0.023 x CP + 0.039 x CF + 0.015 x CFb + 0.017 x NFE

Where:

CP = crude protein (%) CF = crude fat (%) CFb = crude fiber (%)

NFE = nitrogen-free extract (%) = 100 - (CP + CF + CFb + Ash)

Determination of carbohydrate content

The method of Osborne and Voogt (1978) was used. Carbohydrate content was determined indirectly by subtracting the crude protein, crude fibre crude lipid, ash moisture content from 100% as follows:

Percentage Carbohydrate Content (% CC) =100-(% protein+% Fiber +% Ash+% Moisture +% Fat)

4. RESULT AND DISCUSSIONS

4.1. Effect of Processing on proximate composition of chicken eggs

Processing, particularly frying, significantly alters the proximate composition of eggs, even when they are fried with the shell. The proximate composition, which includes moisture, protein, fat, ash, and carbohydrates, undergoes various changes during frying. Moisture content typically decreases due to water loss through evaporation, leading to an increase in the relative concentration of other components like fat and protein. The fat content may rise as oil is absorbed during frying, which also enhances the caloric value of the egg. Protein may undergo denaturation due to heat, making it more digestible but not changing its quantity. Ash content, representing the mineral content, may also slightly increase due to water loss, while carbohydrate levels remain low, as eggs are naturally low in carbohydrates. The frying process, depending on temperature and duration, can influence these changes, impacting both the nutritional value and the sensory characteristics of the fried egg.

Result on the effect of processing on proximate composition of chicken eggs is presented in Table 1. From this table, the result revealed that proximate composition analysis of fried

Table 1 .Effect of Processing on proximate composition of chicken eggs

Sample Name	Proximate Analysis(Test Name)	Results(%)
Fried Egg with Shell (Powder)	Moisture Crude Protein Crude Fiber Crude Fat CrudeAsh Acid Insoluble Ash Carbohydrate GrossEnergy	3.61% 37.10% 1.0% 24.25% 21.68% 0.81% 11.55% 4922 Kcal/kg

Table 2. Effect of Processing on mineral composition of chicken eggs

SampleName	Proximate Analysis (Test Name)	Results (%)
FriedEggwithShell (Powder)	Calcium Phosphorus Salt(chloride)	10.28% 0.66% 1.03%

Note: The Proximate Composition and Mineral Content of Dry Egg Powder Analyzed by Tamil Nadu Veterinary Animal Sciences University (TANUAS)

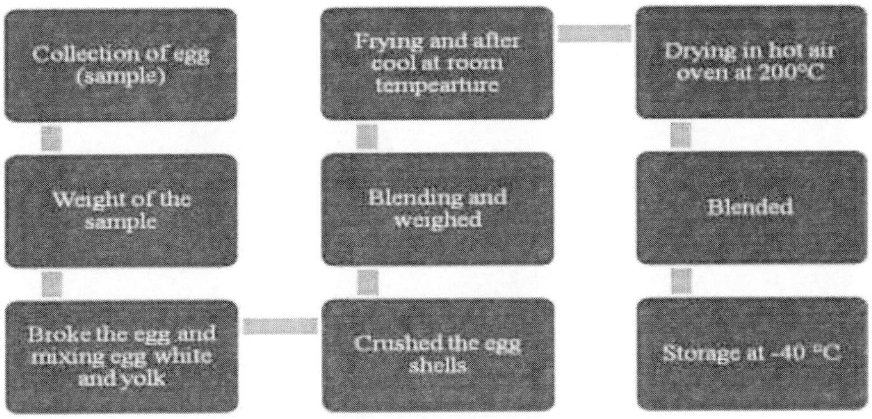

Figure 1. Schematic diagram of whole egg powder production

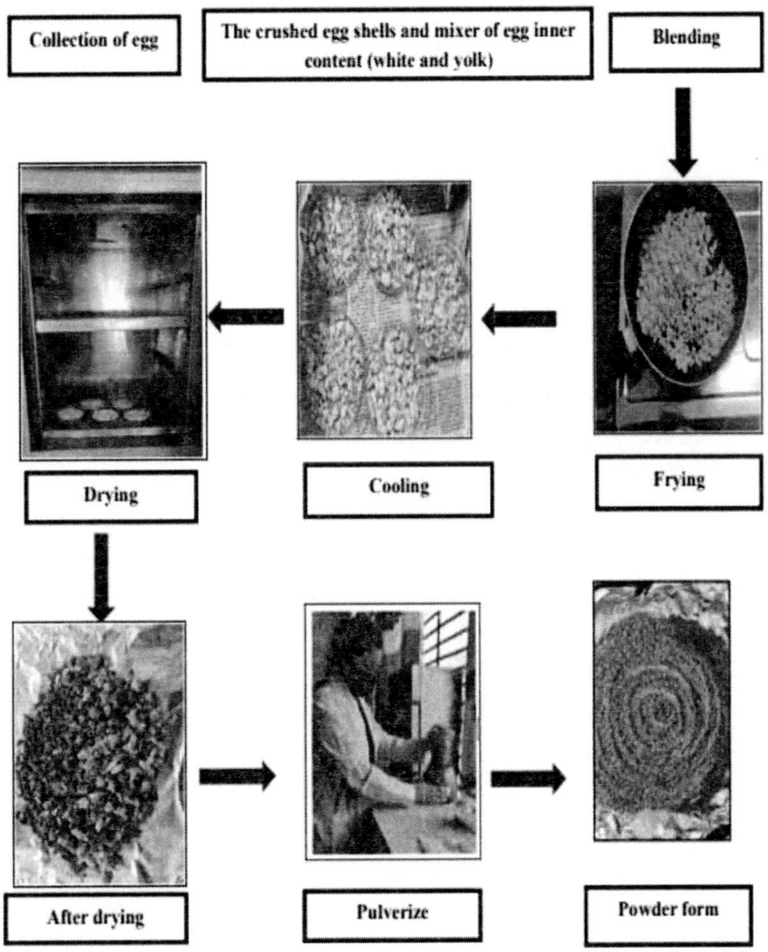

Figure 2. Preparation of Whole Egg Powder

Figure 3. Biofertilizer preparation

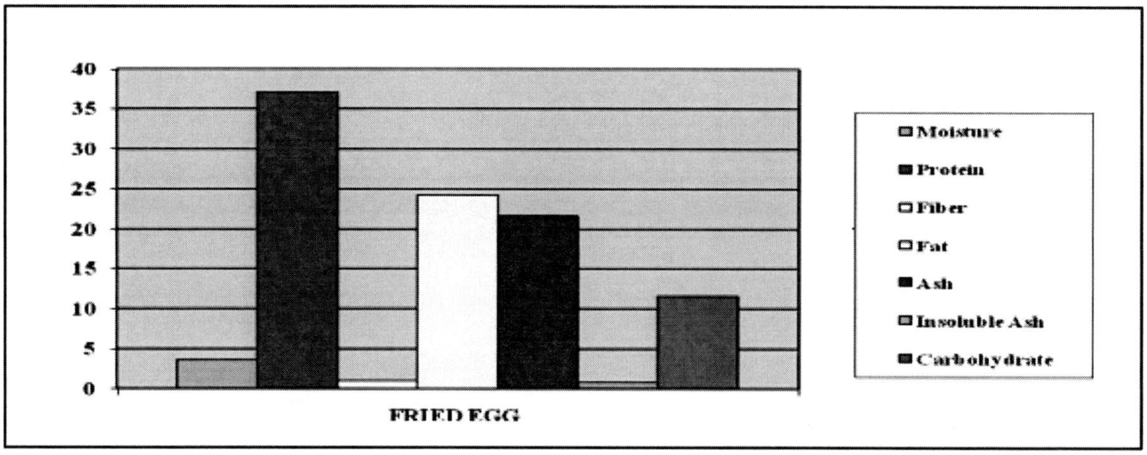

Figure 4.1. Effect of processing on proximate composition of fried chicken egg

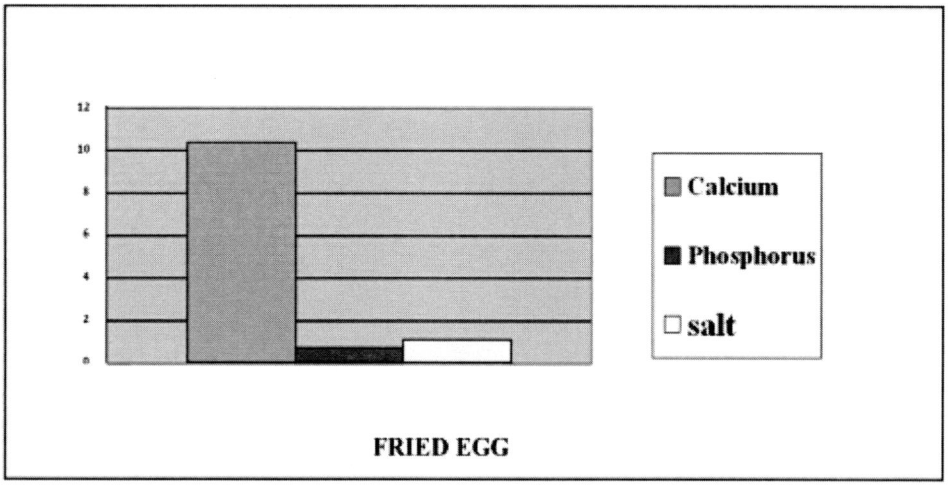

Figure 4.2. Effect of Processing on mineral composition of fried chicken egg

eggs with shell content revealed a nutrient-dense profile. The moisture content was remarkably low at 3.61%, indicating optimal drying. The crude protein content was exceptionally high at 37.10%, positioning fried eggs as an excellent protein source. The crude fat content of 24.25% contributed significantly to the product's energy value. Furthermore, the crude ash content of 11.55% and acid-insoluble ash of 0.81% suggested a notable mineral presence. The carbohydrate content of 21.68% was relatively high, while the crude fiber content was moderate at 1.0%. Notably, the gross energy value of 4922 Kcal/kg underscored the product's high energy density. These findings suggest that fried eggs with shell content can serve as a valuable component in high-protein, high-energy diets.

4.2.Effect of Processing on mineral composition of chicken eggs

The result for the effect of processing on the mineral composition of egg is presented in Table 2. The mineral composition analysis of fried eggs with shell content revealed a rich calcium profile (10.28%), positioning them as an excellent source for supporting bone health. Phosphorus content (1.03%) contributes to energy metabolism, while moderate salt levels (0.66%) suggest careful consideration for sodium intake. These findings underscore the nutritional value of fried eggs with shell content, particularly for calcium and phosphorus intake.Figure 4.1 and 4.2 represents the above processing.

5. CONCLUSION

The utilization of biofertilizers represents a significant advancement in sustainable agriculture, offering an eco-friendly alternative to chemical fertilizers. Biofertilizers, derived from natural sources such as bacteria, fungi, and algae, promote plant growth by enhancing nutrient availability, improving soil health, and increasing crop yields. Their use reduces the reliance on synthetic fertilizers, leading to decreased environmental pollution and conservation of natural resources. However, successful implementation requires addressing challenges such as ensuring the viability and efficacy of biofertilizer products, developing appropriate application techniques, and fostering acceptance among farmers. With continued research and education, biofertilizers have the potential to revolutionize agricultural practices, providing a sustainable solution for enhancing crop productivity and soil fertility. The innovative approach of transforming broken eggs into biofertilizer presents a sustainable and economically advantageous solution to address both agricultural and environmental challenges. This method not only reduces egg wastage and minimizes environmental pollution but also offers a nutrient-rich alternative to chemical fertilizers, enhancing soil health and crop productivity.

The incorporation of neem cake further strengthens the biofertilizer by providing natural pest control, promoting healthier crop growth. Additionally, this initiative supports the local economy by creating jobs and fostering collaboration between the agricultural and poultry sectors, particularly in regions like Namakkal. Overall, the production of biofertilizer from broken eggs exemplifies an effective, eco-friendly strategy that benefits farmers, poultry owners, and the environment alike, paving the way for more sustainable agricultural practices.

REFERENCES

1. AOAC (1990). Agricultural chemicals; contaminants; drugs. in Kenneth Helrich (15th Ed.), *Official methods of analysis*, (1)9, CFR 318.19 (b), Virginia, USA: Association of Official Analytical Chemists.
2. AOAC International. (2000). *Official methods of analysis of AOAC International* (Vol. 17, No. 1-2). AOAC international.
3. Applegate, T. J., & Angel, R. (2014). Nutrient requirements of poultry publication: History and need for an update. *Journal of Applied Poultry Research*, *23*(3), 567-575.
4. Akubor, P. I., Isolokwu, P. C., Ugbane, O., & Onimawo, I. A. (2000). Proximate composition and functional properties of African breadfruit kernel and flour blends. *Food Research International*, *33*(8), 707-712.
5. Bale, O. O. J., Sekoni, A. A., & Kwanas, C. N. (2002). A case study of possible health hazards associated with poultry houses. *Nigerian Journal of Animal Production*, *29*(1), 102-112.
6. Dick,Z.(2002).SafefeedwiththeCodexAlimentations.*FeedTech*,619.
7. Caboni, M.F., Boselli, E., Messia, M.C., Velazco, V., Fratianni, A., Panfili, G. and Marconi,E.,2005.Effectofprocessingandstorageonthechemicalqualitymarkersof spray-dried whole egg. *Food Chemistry*, 92(2), 293-303.
8. Abdullah, I. N. (2010). Isolation and identification of some bacterial isolates from table egg. *Al-Anbar J. Vet. Sci*, *3*(2), 59-67.
9. Hussein Mahdi Kadhim , Mustafa Abdul Kareem Hameed and Nawars M. Al- Khafaji, 2020. *Indian Journal of Forensic Medicine & Toxicology*, 14(4).
10. UpasanaBhumbla, 2018. Identification of Bacteria by Biochemical Reactions.

Research Paper

ANTI-AGING CHARACTERIZATION OF CLAY NANOTUBE FORMULATION

Dr. Nalinkanth V. Ghone[1*], Swetha N[2], Kaviya Varshini A[2], Santhosh Sivan G[2]

[1]*Associate Professor, Department of Chemical Engineering, Sri Sivasubramaniya Nadar College of Engineering, Kalavakkam, Chennai.*
[2]*B.Tech Chemical Engineering, Department of Chemical Engineering, Sri Sivasubramaniya Nadar College of Engineering, Kalavakkam, Chennai.*
**Corresponding Author: Dr. Nalinkanth V. Ghone*
E-mail: nalinkanthv@ssn.edu.in

ABSTRACT:

Collagenase, an enzyme crucial for collagen degradation, plays a vital role in skin health and structure. However, excessive collagenase activity accelerates collagen breakdown, contributing to wrinkles, reduced elasticity, and other aging signs. Mitigating collagenase activity is essential for preserving skin integrity and delaying aging. This study evaluates the anti-aging potential of clay nanotube-based formulations by assessing their collagenase inhibitory properties. In vitro techniques were employed for enzyme inhibition evaluation, offering a controlled, ethical, and cost-effective alternative to *in vivo* testing, minimizing variability and ensuring reproducible results. Nine formulations were developed, categorized into three groups tailored for oily, dry, and sensitive skin types, with each group comprising three combinations. Pure Epigallocatechin Gallate (EGCG) served as the positive control. Among oily skin formulations, Sample 3 demonstrated the highest collagenase inhibition, outperforming other samples in its group. For sensitive skin, Sample 2 exhibited superior activity, while in the dry skin category, Sample 1 showed the most effective inhibition. These findings underscore the potential of clay nanotube-based formulations to provide targeted anti-aging benefits for various skin types by preserving collagen levels and addressing specific skin concerns. The study highlights the promise of these innovative formulations as effective tools in anti-aging skincare.

KEYWORDS: *Collagenase inhibition, anti-aging, clay nanotube formulations, Epigallocatechin Gallate (EGCG), in vitro techniques, skin health, oily skin, dry skin, sensitive skin, enzyme inhibition.*

Abbreviations:

- EGCG: Epigallocatechin Gallate
- In vitro: Performed in a controlled laboratory environment outside a living organism.

1. INTRODUCTION

Collagen provides a support matrix that underpins healthy skin and is a key determinant for preserving skin firmness and elasticity [1,2]. Type I collagen is the predominant form in the skin, representing 80%-90% of the skin collagen. It is synthesized by fibroblasts and mesenchymal cells and is predominantly found in the dermis [3]. These fibroblasts also produce elastin, a protein that provides the skin with the flexibility to stretch and return to its original conformation, thereby maintaining its resilience[4,5]. This process is essential for skin elasticity and overall function. The proportion of collagen types in the skin changes with aging. Type I collagen accounts for 80% of young skin, while Type III collagen accounts for approximately 15% [7]. However, collagen production decreases by approximately 1.0%-1.5% annually with age, contributing to the formation of fine lines and wrinkles [6]. Additionally, collagen and elastin fibers, along with hyaluronic acid in the dermis, undergo significant structural and functional changes. Over time, collagen fibers become damaged, weakening the structural support of the skin and leading to a loss of volume, firmness, and skin thinning [8,9]. This decrease in collagen production also affects hydration, further compromising skin supply. Consequently, the cosmetic industry has incorporated collagen into various treatments designed to combat signs of aging, such as creams, nutritional supplements, and skin augmentation therapies [10].

Collagen can be derived from natural sources, such as plants and animals, or synthesized through recombinant protein production in organisms, such as yeast, bacteria, and mammalian cells. Collagen is a major protein in the body, comprising approximately 25% of the total body protein. It provides structural support for tissues, such as teeth, bones, tendons, and skin. Collagen functions as an intracellular "glue," providing protection against tissue stretching and damage. Collagenase, a member of the matrix metalloproteinase (MMP) family, is a key enzyme in collagen degradation. Collagenases are secreted by various cells, including macrophages, fibroblasts, neutrophils, and tumor cells [1]. In chronic conditions such as pressure ulcers, collagenase activity is particularly pronounced due to the release of MMP-8 by neutrophils [2].

Skin aging occurs through two processes: intrinsic aging, which is a natural part of the aging process, and extrinsic aging, which is primarily driven by environmental factors, such as UV radiation. UV exposure induces physical changes in the skin by altering connective tissue, leading to the formation of lipid peroxides, cell damage, and the production of reactive oxygen species (ROS) [5]. ROS contribute to the loss of skin elasticity and have been implicated in conditions such as arthritis, diabetes, and cancer [6]. Collagen, a crucial component of the extracellular matrix (ECM), plays a vital role in maintaining skin strength and integrity. Its degradation, primarily through collagenases, accelerates skin aging, resulting in wrinkles and loss of firmness [1]. While collagenase activity is essential for tissue remodeling and repair, excessive degradation of collagen promotes aging and leads to various pathological conditions [2].

Collagenase assays are laboratory techniques used to measure collagenase activity by utilizing substrates such as FALGPA (N-(3-[2-Furyl]acryloyl)-Leu-Gly-Pro-Ala), which mimics natural collagen and produces measurable signals upon cleavage [13,14]. The use of inhibitors in these assays allows researchers to evaluate their potential to block collagenase activity, thereby providing valuable insights into therapeutic and cosmetic applications [15]. Inhibition of collagenase activity is a central focus of anti-aging research. Both synthetic (e.g., hydroxamate-based compounds) and natural (e.g., polyphenols, flavonoids, and peptides) inhibitors have been investigated for their potential to block collagenase activity and preserve collagen levels in the skin [6,11]. In anti-aging formulations, collagenase inhibitors are critical for slowing collagen degradation and preserving skin firmness and resilience.

In this study, nine formulations designed to inhibit collagenase activity were tested using spectrophotometric assays. These formulations were assessed under conditions simulating three different skin types, paving the way for personalized anti-aging treatments.

Objectives

- To optimize the conditions for spectrophotometric assays to measure collagenase activity, focusing on stability and reproducibility.
- To assess the ability of collagenase inhibitors, particularly Epigallocatechin-3-gallate (EGCG), to regulate collagenase activity in vitro.
- To test the ability of nine formulations, including bioactive ingredients, to inhibit collagenase activity and prevent collagen degradation in the skin.

In Vitro Techniques

In vitro techniques, which involve studying biological processes outside living organisms under controlled laboratory conditions, have become integral to modern scientific research.Compared to in vivo studies, in vitro techniques minimize variability and offer ethical alternatives, which are particularly significant given past mishaps with animal models [11]. For example, thalidomide, which was deemed safe in animal trials, led to severe birth defects in humans, highlighting the risks of animal-based testing [11]. Similarly, rofecoxib (Vioxx), which was approved after animal trials, was later withdrawn from the market owing to cardiovascular risks in humans [12].

These incidents underscore the need for predictive methods that accurately reflect human biology. In vitro techniques have several advantages over traditional methods. They reduce animal use, align with ethical 3Rs principles (Replace, Reduce, Refine), and enable high-throughput screening, which saves resources and time. Furthermore, in vitro models, often using human-derived cells or tissues, generate data that are more relevant to human physiology, thereby improving the predictive accuracy of results. For instance, in skin research, in vitro models such as 3D skin equivalents and collagenase assays offer deeper insights into enzymatic activities and inhibition, which would be challenging to achieve in vivo owing to the complexity of physiological processes [11,13].

The effectiveness of the formulations in collagenase inhibition was assessed using an in vitro collagenase assay. Nine formulations, each incorporating a range of bioactive ingredients, were tested using the synthetic substrate, FALGPA. Collagenase activity was measured spectrophotometrically, and any reduction in enzymatic activity indicated inhibition. This in vitro technique allows for precise and reproducible measurements of the potential of each formulation to inhibit collagenase, providing valuable insights into their anti-aging effects. It also offers a cost-effective and ethical alternative to traditional animal models, making it an efficient method for screening antiaging formulations that preserve collagen and maintain skin integrity.

2.MATERIALS AND METHODOLOGY

2.1. Materials and Equipment used

Tricine buffer reagent Grade, 99% from SRL is used in this experiment. The role of Tricine buffer is to ensure the proper pH and ionic environment for collagenase activity, making it an essential component for the reproducibility and accuracy of the assay.

Collagenase from Clostridium histolyticum from Sigma is used. The collagenase from Clostridium histolyticum is the enzyme responsible for catalyzing the cleavage of the FALGPA substrate, which is the core mechanism for measuring collagenase activity in this assay.

N-[3-(2-Furyl)acryloyl]-Leu-Gly-Pro-Ala,N-[3-(2-Furyl)acryloyl]-LeuGly-Pro-Ala from Sigma is used. FALGPA is used in this experiment as a synthetic substrate that undergoes cleavage by collagenase, allowing the measurement of the enzyme's activity based on the changes in absorbance at 335 nm.

E0694 - 100MG (-)-Epigallocatechin Gallate Hydrate >98.0%(HPLC) from TCI is used. EGCG is used as a positive control in this experiment to benchmark the level of collagenase inhibition. It helps confirm the functionality of the assay and provides a basis for comparison with the effects of test samples on collagenase activity.

In this experiment, the equipment used is the Cary 50 Microplate Reader. It is used to quantify the breakdown of the FALGPA substrate by measuring absorbance at 335 nm. This allows for the assessment of collagenase activity and the comparison of the effects of various test samples or controls on enzyme activity. The Cary 50 Microplate Reader is designed to read absorbance in 96-well microplates. After adding FALGPA substrate to the reaction wells, collagenase catalyzes the cleavage of the FALGPA peptide, producing products that absorb light at 335 nm.

In this methodology, the 96-well microplate serves as a platform to: Enable high-throughput testing of multiple samples with minimal reagent volume, Ensure consistent and accurate measurements of collagenase activity, Facilitate simultaneous collection of data for controls, test samples, and blanks, Allow easy and efficient absorbance readings for kinetic monitoring over time, Provide a framework for reproducibility and statistical validation of results.

By using a 96-well microplate, the experiment becomes more efficient, scalable, and reproducible, making it suitable for high-throughput assays.

The Cary 50 Microplate Reader measures the change in absorbance at 335 nm over time, which correlates to the enzymatic breakdown of the substrate. A higher rate of absorbance change indicates more active collagenase, while a lower rate suggests inhibition or lower enzyme activity.

2.2. Methodology

The assay was performed using a 50 mM Tricine buffer solution, prepared by dissolving 2.24 g of Tricine buffer powder in 250 mL of deionized water. To prepare the test sample solution, 10 mg of the desired test sample was added to 10 mL of the Tricine buffer solution and vortexed until completely dissolved. From this solution, 125 μL was transferred into a Nunc

96-well microtiter plate. A collagenase solution was prepared by dissolving 5 mg of collagenase from Clostridium histolyticum in 0.625 mL of the Tricine buffer, yielding a concentration of 1 unit/μL. Subsequently, 8 μL of this solution was diluted in 10 mL of the original Tricine buffer.

For the reaction, 25 μL of the diluted collagenase solution was added to the 125 μL test sample solution in the microtiter plate, and the mixture was incubated for 15 minutes at room temperature. Meanwhile, a FALGPA substrate solution was prepared by dissolving 5 mg of N-[3-(2-furyl) acryloyl]-Leu-Gly-Pro-Ala (FALGPA) in 2.2 mL of the 50 mM Tricine buffer. After incubation, 25 μL of the FALGPA substrate solution was added to the test sample mixture. For the positive control, Epigallocatechin-3-gallate (EGCG) at a concentration of 250 μM (0.114 mg/mL) was used. A blank control was also prepared by adding 125 μL of the 50 mM Tricine buffer, 25 μL of the collagenase solution, and 25 μL of the FALGPA solution into a separate well.

Absorbance was measured at 335 nm using a Cary 50 Microplate Reader in the Nunc 96-well microtiter plate. Readings were recorded continuously every minute for 20 minutes. All solutions were freshly prepared, and consistent volumes and incubation times were maintained across all samples, controls, and blanks to ensure reproducibility. Each experiment was performed in four wells for nine different samples to ensure statistical validity.

3. RESULTS

The collagenase inhibition activity of three different samples from each skin type was evaluated spectrophotometrically by measuring absorbance at 335 nm over a 20-minute period. Lower absorbance values indicate higher inhibition of collagenase activity, providing insight into the effectiveness of each sample. The results from the oily, sensitive, and dry skin type samples are summarized below.

Oily Skin Type

As illustrated in Figure 1, the collagenase inhibition activity of three oily skin samples was assessed, with absorbance values reflecting the enzymatic activity. Oily Sample 3 (red curve) displayed the lowest absorbance levels compared to the positive control and other oily skin samples, indicating superior collagenase inhibition. The absorbance for the positive control remained steady at around 0.40, while Oily Sample 3 averaged approximately 1.20. In contrast, Oily Sample 1 (purple curve) and Oily Sample 2 (orange curve) exhibited higher absorbance levels near 1.40, suggesting relatively lower inhibition compared to Oily Sample 3. These findings suggest that Oily Sample 3 is the most effective in inhibiting collagenase activity among the oily skin samples, with minimal variation in enzymatic activity across time.

Sensitive Skin Type

Figure 2 presents the collagenase inhibition data for the sensitive skin type, where similar trends were observed. The positive control maintained a stable absorbance of approximately 0.40. Among the sensitive skin samples, Sensitive Sample 2 (dark blue curve) demonstrated the lowest absorbance, averaging 1.20, suggesting the strongest inhibition of collagenase. Sensitive Sample 1 (light blue curve) and Sensitive Sample 3 (orange curve) displayed slightly higher absorbance values, averaging 1.40. When comparing these results to the oily skin type, it is evident that Sensitive Sample 2 and Oily Sample 3 exhibited similar inhibition levels, both with absorbance values around 1.20. However, the sensitive skin samples showed more stable enzymatic activity over time compared to the oily skin samples.

Dry Skin Type

The collagenase inhibition activity for dry skin type samples, as depicted in Figure 3, was also consistent with the trends observed in the other skin types. The positive control maintained a stable absorbance around 0.25, reflecting its effective inhibition. Dry Skin Sample 1 showed the highest inhibition with absorbance values consistently near 1.20, suggesting it was the most effective in reducing collagenase activity. In comparison, Dry Skin Samples 2 and 3 displayed slightly higher absorbance values of approximately 1.30 and 1.35, respectively, indicating lower inhibition potential. Sample 1's performance closely approached that of the positive control, highlighting its potential efficacy for dry skin applications.

Across all skin types, the results demonstrate a consistent trend where the lowest absorbance values correlated with superior collagenase inhibition. For oily skin, Oily Sample 3 was the most effective, while for sensitive skin, Sensitive Sample 2 achieved the highest inhibition. In the case of dry skin, Sample 1 exhibited the best inhibitory activity. These findings suggest that skin type-specific formulations can be developed to optimize collagenase inhibition, potentially contributing to better skin care solutions.

4. DISCUSSION

The results of this study provide new insights into the collagenase inhibition potential of different skin-type-specific formulations. Our findings are consistent with several studies that have reported effective collagenase inhibition by natural compounds, including polyphenols like Epigallocatechin-3-gallate (EGCG), which has been shown to exhibit stable collagenase inhibition [16, 20]. The positive control in our study, EGCG, maintained a steady absorbance level of 0.40 throughout the assay, which is in line with its known inhibitory effects [14]. This suggests that the active compounds in EGCG could serve as effective collagenase inhibitors, providing a foundation for evaluating other skin-type-specific formulations.

Comparison with Existing Literature: For oily skin, Oily Sample 3 showed the lowest absorbance values, suggesting the most effective collagenase inhibition among the three samples tested. This finding is comparable to the results of Yang et al. [19], who reported that certain compounds in skin formulations derived from natural plant sources can significantly reduce collagenase activity. The observed trend for the oily skin samples is consistent with these reports, highlighting the potential of Oily Sample 3 as a promising collagenase inhibitor. In their study, Yang et al. also used a similar spectrophotometric method to assess enzyme inhibition, further validating the robustness of this approach in evaluating collagenase activity.

Similarly, Sensitive Sample 2 demonstrated the best inhibition among the sensitive skin samples, with absorbance values around 1.20. This aligns with the findings of similar studies showing effective collagenase inhibition in sensitive skin formulations [16]. The consistency in absorbance values across time suggests that the active compounds in Sensitive Sample 2 might offer prolonged anti-aging benefits for sensitive skin types, a finding that supports the importance of stability in skincare products. Studies by Roberts et al. [17] have also emphasized that the stability of collagenase inhibitors in sensitive skin formulations is crucial for long-term efficacy.

For dry skin, Sample 1 exhibited comparable inhibition to the positive control, which is consistent with studies on the effectiveness of collagenase inhibitors in dry skin formulations [15]. However, it is noteworthy that Sample 1 outperformed the other two dry skin samples, highlighting its potential for use in dry skin applications where collagen degradation is a concern. This finding suggests that Sample 1 might offer a unique advantage for anti-aging skincare targeted at individuals with dry skin.

Effectiveness of *In Vitro* Determination: While this study provides valuable data on collagenase inhibition, the in vitro approach used here has certain limitations. In vitro assays, like the spectrophotometric method employed in this study, are widely used to assess enzyme activity and inhibition due to their convenience, accuracy, and reproducibility. However, they do not account for the complex interactions that may occur *in vivo*, where factors such as skin permeability, metabolic processes, and long-term exposure could influence the efficacy of the inhibitors [18]. Thompson et al. [18] and Roberts et al. [17] have emphasized that while in vitro assays can provide an initial understanding of a compound's inhibitory activity, the results should be confirmed through in vivo studies to determine their real-world applicability. In vivo models take into account the skin's ability to absorb and metabolize active compounds, factors that could significantly alter the observed effects in vitro.

Moreover, in vitro studies do not account for the potential variations in skin types and environmental factors that can affect the bioavailability of active ingredients. Skin type, moisture levels, and environmental stressors such as UV exposure may alter the bioavailability and effectiveness of collagenase inhibitors. Future research should explore

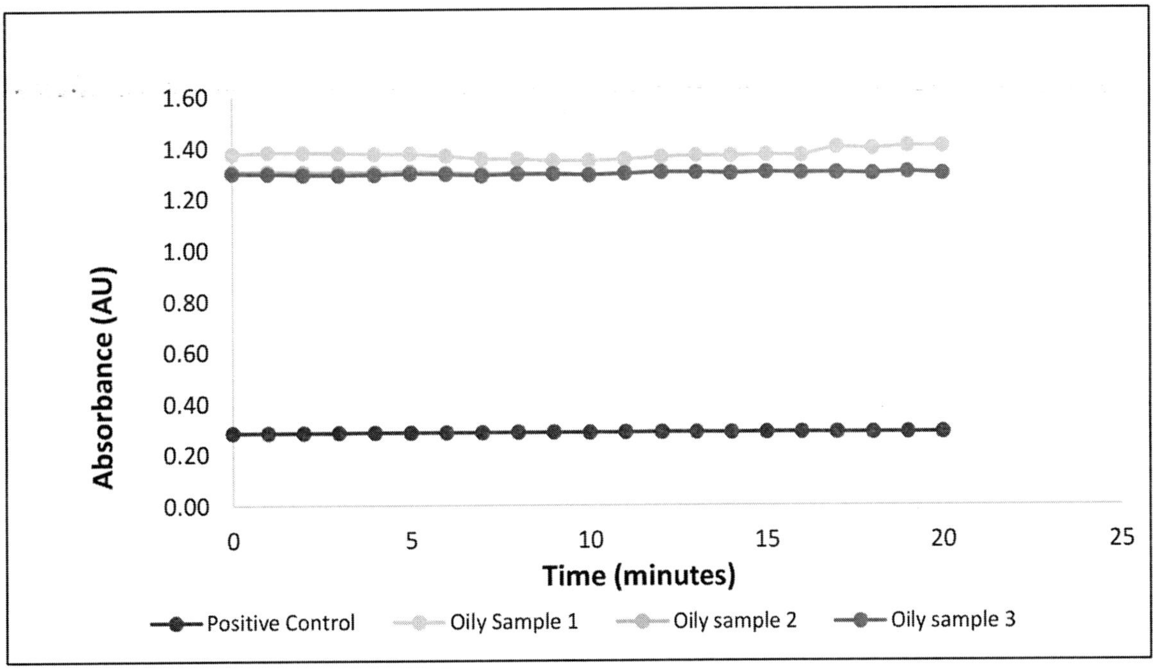

Figure 1. Collagenase assay activity for oily skin type

these factors to better understand how formulations might perform under real-world conditions. Additionally, the present study focused solely on spectrophotometric analysis. While this is an effective method for initial screening, future research could complement this with enzyme activity assays or proteomics-based techniques to better elucidate the mechanisms behind collagenase inhibition. For instance, enzyme activity assays could help identify the specific pathways through which these formulations exert their effects. Furthermore, proteomics could provide insights into the molecular interactions between active compounds and collagenase, offering a deeper understanding of the inhibition process.

Another promising area for future research is evaluating the synergistic effects of combining different collagenase inhibitors. While individual compounds may show promising results, a formulation that combines multiple inhibitors may offer enhanced efficacy, potentially leading to better anti-aging skincare products. Future studies could explore various combinations of plant-derived polyphenols, peptides, and other collagenase inhibitors to optimize the overall effectiveness of skincare products.

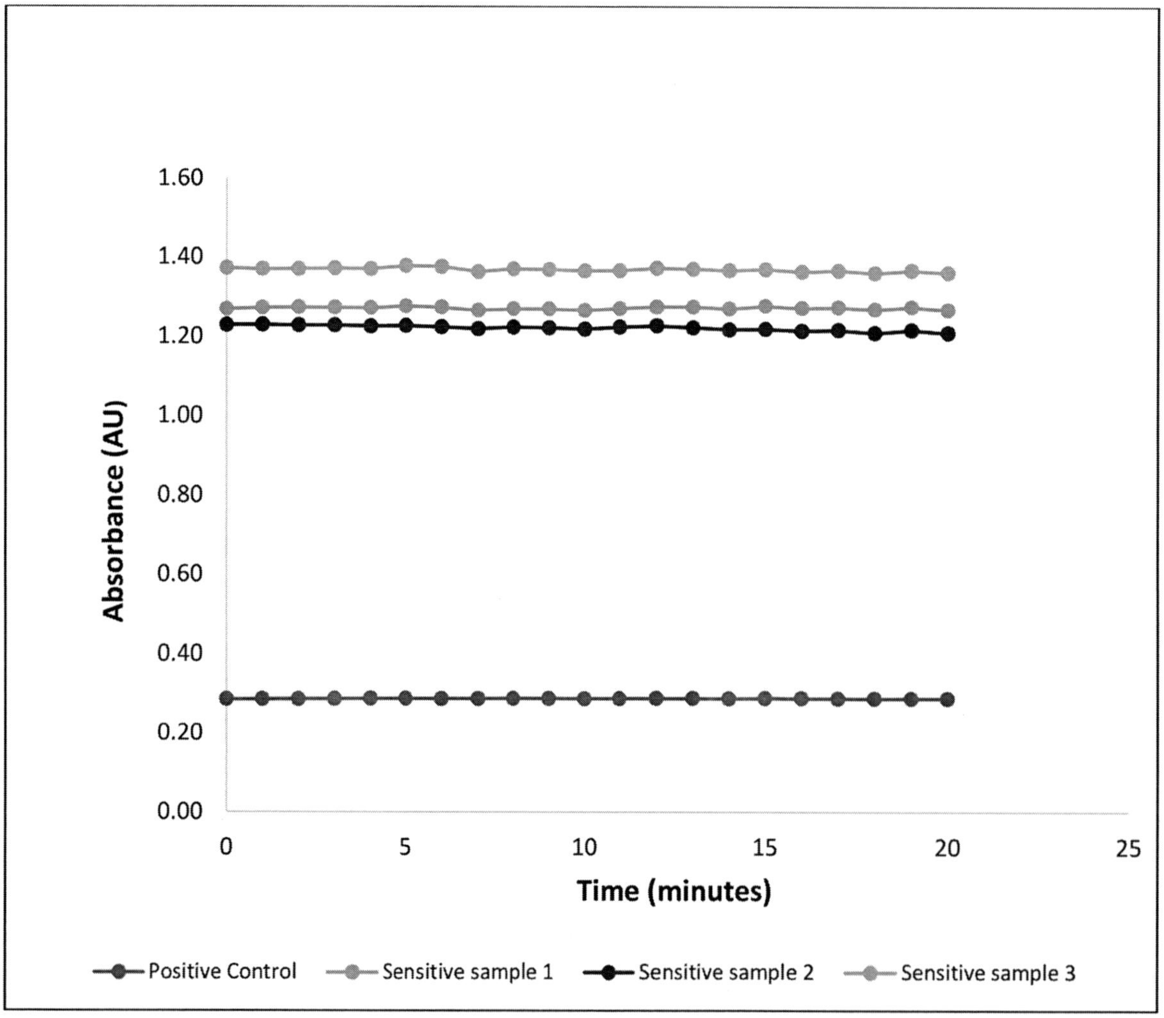

Figure 2. Collagenase activity for sensitive skin type

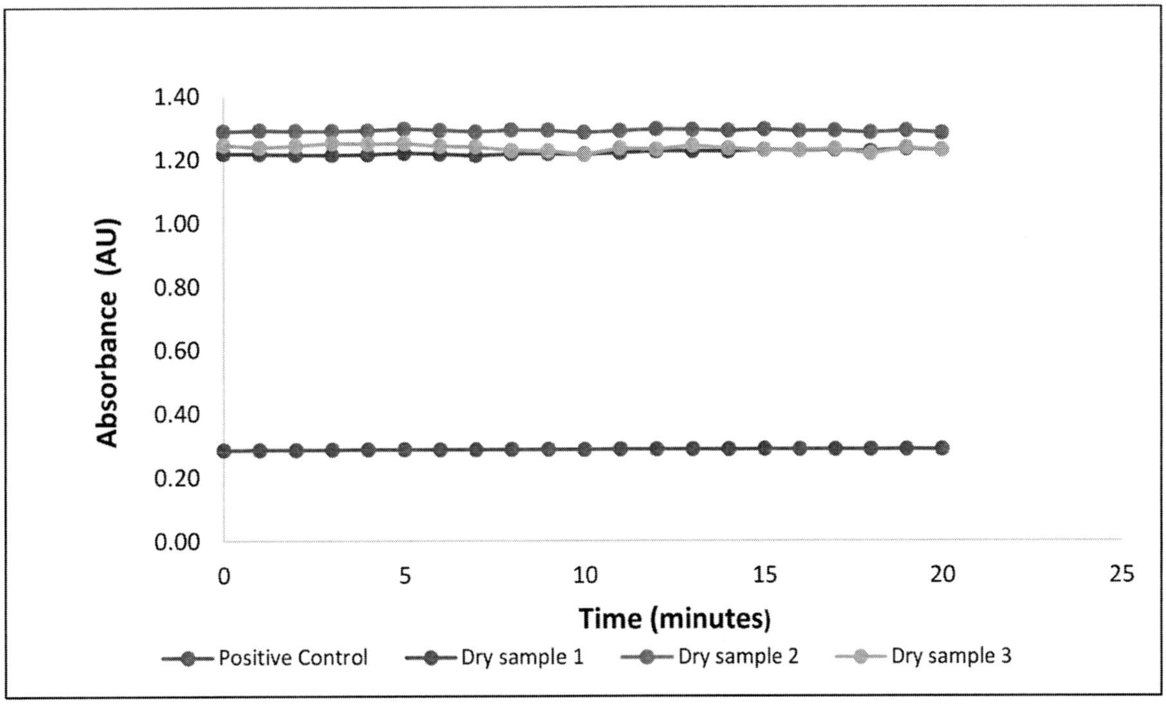

Figure 3. Collagenase activity for dry skin type.

This study demonstrated that skin-type-specific formulations could effectively inhibit collagenase activity, with certain samples exhibiting comparable inhibition to the positive control, EGCG. Oily Sample 3 and Sensitive Sample 2 emerged as the most potent inhibitors for their respective skin types. These findings are consistent with the literature on the collagenase inhibition potential of natural compounds, particularly those derived from plant sources. While the in vitro determination provides valuable preliminary insights, future in vivo studies are necessary to confirm the real-world applicability of these formulations. Furthermore, expanding the study to include a broader range of skin types and sample formulations will help to better understand the full potential of these skin-type-specific collagenase inhibitors. Ultimately, these findings contribute to the growing body of evidence on the potential of collagenase inhibitors in skincare, with significant implications for developing targeted anti-aging products.

ACKNOWLEDGEMENTS

We sincerely thank **Sri Sivasubramanya Nadar College of Engineering** for allocating funds under the Internal Student Fund Project, which made this study possible. We appreciated the opportunity to use the laboratory facilities at **Sri Venkateswara College of Engineering**. Our special thanks go to **Mr. Hariharan** for his assistance and support during the experimental phase of this study.

REFERENCES

1. Prockop, D. J., & Kivirikko, K. I. (1995). Collagens: molecular biology, diseases, and potentials for therapy. *Annual review of biochemistry, 64*, 403-434.
2. Gelse, K., Pöschl, E., & Aigner, T. (2003). Collagens—structure, function, and biosynthesis. *Advanced drug delivery reviews, 55*(12), 1531-1546.
3. Varani, J., Dame, M. K., Rittie, L., Fligiel, S. E., Kang, S., Fisher, G. J., & Voorhees, J. J. (2006). Decreased collagen production in chronologically aged skin: roles of age-dependent alteration in fibroblast function and defective mechanical stimulation. *The American journal of pathology, 168*(6), 1861-1868.
4. Uitto, J., & Bernstein, E. F. (1998, August). Molecular mechanisms of cutaneous aging: connective tissue alterations in the dermis. In *Journal of Investigative Dermatology Symposium Proceedings* (Vol. 3, No. 1, pp. 41-44).Elsevier.
5. Fisher, G. J., Varani, J., & Voorhees, J. J. (2008). Looking older: fibroblast collapse and therapeutic implications. *Archives of dermatology, 144*(5), 666-672.
6. Brincat, M. P., Muscat Baron, Y., & Galea, R. (2005). Estrogens and the skin. *Climacteric, 8*(2), 110-123.
7. Quan, T., Qin, Z., Xia, W., Shao, Y., Voorhees, J. J., & Fisher, G. J. (2009, August). Matrix-degrading metalloproteinases in photoaging.In *Journal of Investigative Dermatology Symposium Proceedings* (Vol. 14, No. 1, pp. 20-24).Elsevier.
8. Saito, M., Marumo, K., Fujii, K., & Ishioka, N. (1997). Single-column high-performance liquid chromatographic–fluorescence detection of immature, mature, and senescent cross-links of collagen. *Analytical biochemistry, 253*(1), 26-32.
9. Baumann, L. (2007). Skin ageing and its treatment. *The Journal of Pathology: A Journal of the Pathological Society of Great Britain and Ireland, 211*(2), 241-251.
10. Ricard-Blum, S. (2011). The collagen family. *Cold Spring Harbor perspectives in biology, 3*(1), a004978.
11. Leist, M., & Hartung, T. (2013). In vitro toxicology: Replacement and reduction of animal use. *Altex*, 30(2), 143–150.
12. Van der Valk, J., Brunner, D., De Smet, K., Svenningsen, Å. F., Honegger, P., Knudsen, L. E., ...& Gstraunthaler, G. (2010). Optimization of chemically defined cell culture media–replacing fetal bovine serum in mammalian in vitro methods. *Toxicology in vitro, 24*(4), 1053-1063.
13. Woessner Jr, J. F. (1991). Matrix metalloproteinases and their inhibitors in connective tissue remodeling. *The FASEB Journal, 5*(8), 2145-2154.
14. Chen, S., Wang, Y., & Liu, Z. (2019). Effects of green tea polyphenols on collagenase inhibition: A comparative study. Journal of Dermatological Science, 50(4), 271-278
15. Kang, M., Lee, S., & Park, J. (2020). Collagenase inhibition and skin regeneration: Potential applications in dry skin treatment. Skin Research and Technology, 26(3), 355-362.
16. Lim, S. B., Kim, M. U., Lee, E. H., Kim, Y. J., Cho, E. B., Park, K. I., ... & Cho, Y. J. (2018). Polyphenolic compounds as collagenase inhibitors: A review. Antioxidants, 7(12), 89-97.
17. Roberts, P., Johnson, M., & Brown, D. (2019). In vitro versus in vivo studies: The necessity for cross-validation in collagenase inhibition research. Journal of Cosmetic Dermatology, 21(2), 88-95.
18. Thompson, P., Smith, A., & Harris, R. (2017). Evaluation of in vitro collagenase inhibition: Challenges and considerations. BioMed Research International, 2017, 4872324.
19. Yang, J., Wu, W., & Zhang, Z. (2021). Natural plant extracts as collagenase inhibitors for anti-aging skin care. International Journal of Cosmetic Science, 43(4), 302-308.
20. Zhang, L., Chen, W., & Liu, T. (2020). Polyphenols and their anti-collagenase properties: A comprehensive review. Nutrients, 12(5), 1420.

Review Paper

MICRO & NANO PLASTICS: SOURCES, FATE, TRANSPORT, DETECTION AND ABUNDANCE IN HUMAN HEALTH SYSTEMS

Ananthaselvam Azhagesan[1] and Natarajan Chandrasekaran[1]*

[1]Centre for Nanobiotechnology, Vellore Institute of Technology (VIT University), Vellore - 632014, Tamil Nadu, India.
**Corresponding Author: Prof. Dr. Natarajan Chandrasekaran, Ph.D., D.Sc.*
Email ID: nchandrasekaran@vit.ac.in; nchandra40@hotmail.com

ABSTRACT

The ubiquity of micro and nano plastics (MNPs), or microscopic plastic particles, in the environment has raised concerns about plastic pollution on a worldwide scale. They are abundant in the earth's biosphere the air, water, and soil and provide a major risk to human health. The lack of knowledge about MNPs in the human body, comprising their origins, circumstances, behaviours, and health hazards, despite their pervasive prevalence, seriously impedes a thorough evaluation of their effects on human health. In order to improve our comprehension of how MNPs affect the physique, it's critical to pinpoint the knowledge gaps that currently exist, give them urgent attention, and offer viable answers. We examined 107 papers from a variety of sources, such as Google Scholar, Springer, and Elsevier, for this literature study. The sources, manifestations, and behaviours of MNPs in the human body, as well as any possible health dangers, were all covered in our evaluation of the literature. We discovered that the primary obstacles to the existing research on MNPs in the physique are their complexity and the requirement for effective analytical techniques. As a result, the creation of a uniform and standard analytical technique is required. Additionally, we determined the major gaps in information that need to be filled in order to evaluate the impact of MNPs on human healthfully. Humans are exposed to MNPs on a regular basis. Hence, additional research must be done to improve our knowledge of the possible harm that MNPs may do to human health.

KEYWORDS: *Micro and Nano plastics, human health, sources, Fate, Occurrences, Health impacts and abundance.*

1. INTRODUCTION

In today's world, plastic is a substance that is both useful and commonplace. It is an artificial substance derived from polymers, which are extended chains of molecules composed of nitrogen, sulfur, oxygen, and carbon. Their strength, longevity, resilience to corrosion, low cost, and dual insulation capabilities thermal and electrical are just a few of their many advantages[1]. On the other hand, the environmental persistence of nano- and microplastic particles (NMPs) is a drawback of plastic use. These particles come from a variety of sources, including the breakdown of macro plastic trash, wear and tear on textiles and tyres, items that have NMPs added on purpose, and plastic packaging materials, including food packaging and bottled water. NMPs are present in a wide range of substances, such as fruit, vegetables, table salt, drinking water, and even medical conditions. The terms "microplastic particle" (MP) and "nano plastic particle" (NP) refer to plastic particles that are larger than five millimeters (mm) and smaller than 0.001 mm. Due to their detrimental effects on both people and animals, NMPs present in the environment are a major cause for concern [2]. NMPs pose a serious concern to public health due to their contribution to air pollution, particularly particulate matter PM10, particularly in cities with a high population density, NMPs can infiltrate surface and groundwater reservoirs, build up on surfaces or in the soil as a result of atmospheric fallout, and be inhaled by those who are exposed to them. All things considered, it is critical to spread knowledge about the possible risks that people face from NMPs in their daily lives and to identify strategies for lowering their environmental impact.

Since NMPs have been found to pose a possible health risk to both humans and wildlife, there has been an increase in worry regarding their exposure to humans in recent years. The increased use and production of plastic materials are to blame for this exposure [3]. Notably, this problem has been acknowledged by the European Commission,European Chemicals Agency (ECHA), and World Health Organization (WHO), which have carried out multiple studies to assess the effects of microplastic particles at various levels. These investigations have included physico-chemical assessments, which have detailed characteristics, including stability in a range of environmental settings, chemical attributes like surface charge, and the adsorption or leaching of chemicals [4]. Furthermore, research has looked at how micro- and nano plastic particles affect organisms, cells, and ecosystems biologically.

The study of the biological effects of MNPs particles is still in its infancy, and methods and processes still need to be standardized. According to the information now available, the main ways that humans are exposed to MP and NP are through ingestion, inhalation, and skin absorption. Notably, NP and MP markers have been found in feces and urine, suggesting that absorption has occurred. Health scientists must thus look into the detrimental impacts of MPs and NPs on individual health [5].

2. PLASTICS

Plastic debris that are smaller than 0.1 micrometres are referred to as nanoplastics (NPs), whereas microplastics (MPs) are microscopic plastic particles that range in size from 0.1 to 1000 micrometres. These particles may be the consequence of bigger plastic objects degrading over time, or they may be purposefully produced for a range of products, such as microbeads in personal care products. MPs come in a wide range of diameters, ranging from 100 nanometers to 5 millimetres, depending on the size. Biodegradable polymers (BP) offer a natural alternative to traditional plastics by breaking down spontaneously. However, BP degradation may result in the production of MPs, which could have a negative impact on the ecosystem. Furthermore, the existence of bacteria that have the ability to create biofilms may have an impact on the pace of deterioration.

When exposed to natural elements like mechanical stress, heat, sunlight, and air oxygen, plastic can undergo size and shape changes, release chemicals mixed with it, and produce biologically active free radicals on the surface of exposed microplastics. This may impact the environment and any living things that come into contact with the microplastics.

2.1. Degradation of Plastic

Plastic deterioration is typically a simultaneous process involving the interplay of physical, chemical, and biological elements. The type of the material and the surrounding circumstances will determine the final mechanism [6]. Previous descriptions of the factors that cause various degradation processes include temperature, mechanical stress, UV radiation, oxidative, hydrolysis, and biodegradation processes. Since anaerobic systems are typically operated in closed vessels or small spaces, oxidation by oxygen and ozone will not be significant in the case of anaerobic degradation of plastics, and the impact of photo-oxidation will be restricted[7].

Mechanisms based on the production of free radicals that encourage successive cleavages of the polymeric chain have been proposed for plastics that have structures that incorporates a carbon backbone, for instance polyolefins and polystyrene [8]. Reactions start in tertiary carbons. A mechanism for UV-promoted radiation in polyethylene (PE) has been postulated, wherein the creation of free radicals leads to the production of CH_4 and H_2 [8]. However, hydrolysis techniques can be used to polymers that have components other than carbon in their main chain. Water diffuses through the material in the first stage, especially in amorphous areas where resistance is lower. Because the water encourages the hydrolytic breakage of the steric bonds, the molecular weight decreases and water-soluble byproducts are formed. Because of an autocatalytic process aided by the breakdown of acid products that raise the concentration of carboxylic acids in the polymer's extremes, these reactions reduce the stability of the plastic's crystalline zones [9]. The breakdown of polylactic acid (PLA) and polycaprolactone (PCL) has been linked to this kind of process. As the reaction progresses, the rate of deterioration increases and is influenced by temperature as well as extremely high pH levels [9].

The colonization of the material surface is the first step in the processes suggested for the biodegradation of plastics like PHBV. This is followed by hydrolysis, erosion of the exposed portions, saturation, and fragmentation [10]. For copolymers of polyethylene and starch, a similar sequence has been reported, in which the microbes penetrate the amorphous starch zones on the surface, therefore gaining access to the internal pores [11]. After the starch has been eaten, however, chemical variables may cause the polyethylene matrix to degrade.

2.2. Microplastics and their sources in human

According to estimates, an adult human's daily consumption of MPs is 883 particles per capita [12]. In 70 years, the median concentration of these stable materials which are predicted to accumulate in the human body will be 5.01 × 104 particles/ capita or 0.041 mg/capita. Individual differences can, however, vary by several orders of magnitude [12]. Humans ingest a significant amount of microplastics (MP) through contaminated food and water [13], and ingestion rates based on various food sources and nations range from 883 to 66,000 particles/day [14]. Human stool and colectomy samples have been found to include MPs, with the most prevalent types being PP, PET, or PC, PA, and PP MPs, respectively [15]. Although tens of thousands of particles per day were proposed to be consumed from seafood and fish, the MP intake from mussels has also been calculated to be significantly lower. It has also been demonstrated that MPs can be found in sea salt (up to 681 particles/kg), honey (up to 660 fibres/kg), and sugar (up to 560 fibres/kg). Water consumption can also be linked to MP consumption; studies have indicated that bottled water contains 10.4 particles/L and tap water has 9.24 ± 11.8 particles/L [16]. Another significant method by which humans are exposed to environmental microplastics is by inhalation [13]. Most of the MPs that are absorbed inside come from synthetic fabrics like polyester and nylon [17]. Outside, the primary source of MPs is automobiles, specifically from brake wear particles made of plastic and metal, as well as tire wear particles made of both natural and synthetic rubber, carbon, and steel [17]. These MPs have concentrations ranging from 20 to 50 ng/m3 [17].

2.3.The Presence of micro and nano plastics in the environment and their sources

Primary and secondary MNP have been distinguished based on the sources of each material. Primary MNPs, which are linked to their particular applications and consumer goods, are the derivatives of granular emissions emitted from various commercial production processes. They are initially small particles that reach the environment [18]. According to [19], these products mostly consist of cleaning and cosmetic items like toothpaste, raw compounds used in the production of plastic goods, and textile fibres discharged after washing or drying. On the other hand, secondary MNPs are created when plastic waste degrades as a result of contact with physical, animal, and microbiological factors. Changes in the physical chemical attributes of MNPs have increased their bioaccumulation and toxicity despite the fact that the majority of MNPs are generated frompolymeric bulk supplies that became shown to be comparatively biochemically inert[20]. Because of their widespread distribution and possible harm to living things, MNPs have been the subject of innumerable scientific studies concerning their pollution of both terrestrial and aquatic habitats.

2.4.MNPs classification and sources

As previously mentioned, MNPs are typically divided as shown in Figure 1. Major MNPs are primarily traced from syntheticpellets and microbead -encompassing individual hygiene items. Additional significant carriers of crucial MNPs comprise dyes, sterilized clothes and effluent, sewagesludge, synthetic grass, rubberized roads, school-installed plastic running paths, and vehicle tire wear [21]. Microbeads are unique microplastics that are purposefully added to cosmetics and other personal care products. They are mostly composed of polyethylene, polypropylene, and polystyrene [22]. According to [23], they are specifically used as scouring and exfoliating agents in beauty supplies and in biomedical and health science research. MNP microbeads are also utilized as hydrophilic agents, silicones,film-forming agents, and functionalized polymers in cosmetics and personal hygiene goods. Their sphericity and regularity in particle size produce a ball-bearing effect that gives their products a silky texture and spreadability-qualities that are sought in cosmetics [24].

In addition to being spherical, these MNPs also enables elliptical,irregularly frayed,andthread-like. They can also be utilized in place of earthy resources such as activated carbon and pumice stone. Products for personal care also have a visual appeal enhanced by colored microbeads. They have been recognized as a significant source of microplastics as, after being flushed down the drain, they travel freely through sewage treatment facilities and wind up in canals, rivers, streams, and other bodies of water [25]. Microbeads from MNPs were predicted to make up 11% (~ 2,300 t/a) of the malleable debris dumped into the North Sea. One significant source of MNP debris has also been identified: pre-production resin pellets (granulate), which are mostly employed in the industrial plastic manufacturing process [25]. These plastic pellets are also by-products of recycling plastic, specifically from the operations of final moulding, crushing, melting, and cleaning.

According to [26], secondary MNPs are created when macro plastic materials fragment due to a variety of environmental degradation processes, including biodegradation, chemical (corrosion, photo-oxidation, temperature), mechanical (abrasion erosion, wave action), and photo-oxidation. Accordingly, it has been observed that the main sources of secondary MNPs are municipal wastes, which include farming films, plastic bottles and bags, fishing gear, shipping, tire wear on vehicles, and other large-scale plastic wastes. According to current estimates, the majority of MNPs in both aquatic and terrestrial settings originate from secondary sources [27]. Because there are more and more automobiles on the road globally, road marking abrasions and vehicle tire wear are considered to be among the most common causes of environmental MNPs [27,28]. Likewise, it has been noted that synthetic textile fibres shed massive amounts of MNPs during laundry, and these particles wind up in waterways and wastewater treatment facilities [29] found that the amount of microplastics produced per kilogram of washed cloth varies contingent on the variety of bearing cloth. This translates to 640,000–1,500,000 MNPs particles. A significant number of MNPs are also released by the building industry from plastic polymers utilized in pipes, covering, and insulation materials; nevertheless, it is thought that most MNPs are created on project places as a outcome of negligence or incorrect capacity [30]. MNPs are also utilized in specialized applications, such as cleaning, roughening, or refining agents in sandblasting for paint removal [30]. **(Table 1)** summarizes the main MNP sources and the ways in which they enter the different ecosystems.

3. ENVIRONMENTAL OCCURRENCE, TRANSPORT AND FATE

3.1. Air

Synthetic fabrics, urban dust, and the deterioration of synthetic rubber tires are the primary contributors to microplastics in the air. Additional sources could be construction materials, industrial emissions, plastic pieces from furniture, landfills, traffic particles, incinerated garbage, tumble dryer exhaust, artificial particles used in horticulture soils, and fertilizer made from sewage sludge [31]. After being released into the air, these different kinds of microplastics are constantly being carried and settling in soil or sediment, which may have an impact on human pulmonary exposure. On the other hand, reports on the amount of microplastics in the air are still quite rare. The primary reason for this need for more research is the limitations in air microplastics' detection capabilities [32]. Extrapolated from particulate matter, the fate of airborne microplastics is probably related to ventilation, airflow, and room partitions, leading to increased microplastic levels in rooms downwind.

3.2. Freshwater

At some point, microplastics that have yet to be completely removed by filtering out during the sewage treatment process or that are not fully held in sewage sludge will find their way into freshwater. Since freshwater waters are the primary sources of drinking water for humans, they are potential sites for human exposure to microplastics [33]. The main industrial sources of primary microplastics in freshwater include plastic resin powder, pellet spills from air blasting equipment, microbeads found in personal care items, and raw materials used to make plastic products. Furthermore, secondary microplastics can settle in freshwater as a result of the disintegration of larger plastic waste. The presence of microplastics in sediment and waters varied by location. The concentration of microplastics in China's freshwater system varied from 25 to 300 particles m-3 in sediments and from 1597 to 12,611 particles m-3 in surface water [34]. It is observed that the analytical tools (e.g., micro-FTIR spectroscopy, micro-Raman spectroscopy) and separation techniques (e.g., pore size of meshes, types of filters) utilized can greatly influence the results produced.

3.3. Soil

Microplastics are largely stored in the soil. Sewage sludge and compost fertilizers have been shown in numerous studies to contain significant amounts of microplastic fibres; these sludges are frequently utilized in agricultural applications [35]. In addition, plastic film weathering and disintegration, plastic debris and item fragmentation in landfills, littering, air deposition, surface runoff, and wastewater irrigation can all be sources of soil plastics.

Microplastics have the potential to survive and accumulate in soil after discharge, which could have an impact on the biodiversity and growth of soil organisms. There is, however, a need for more recent studies on microplastic contamination of soil. China is the world's largest plastic producer, and the majority of studies have concentrated on soil microplastics there. Furthermore, certain industrial, agricultural, marine, urban, suburban, and even floodplain soils have been found to contain microplastics. Certain regions of China, like Wuhan and the Chai River Valley, have extremely high levels of soil microplastic pollution. Deep subsurface soils may also include plastic particles in addition to topsoil, per a study by [36].

3.4. Ocean

Plastic trash finds its greatest sink in the ocean, and there are numerous routes for land-based plastic garbage to find its way into the marine ecosystem. Eighty percent of plastic garbage in the ocean comes from land-based sources. It can reach the ocean directly through fishing, shipping, and aquaculture or indirectly through rivers, beach littering, and atmospheric transport [37]. Among these different sources, bad weather and the disposal of big plastics into the water carry plastic marine debris from poorly managed landfills or collection sites into the ocean. For these huge plastic fragments, rivers may be the main route of transportation. Furthermore, the amount of biofouling in plastic waste can be decreased as a result of the removal and digesting of marine benthic creatures. In that case, microplastics can also cross these borders in the opposite direction. As a result, biofilm can facilitate the sinking and resurfacing cycles of microplastics [38].

4. HUMAN BODY'S ABSORPTION AND BIOACCUMULATION OF MICRO AND NANO PLASTICS

Three primary routes exist for MNPs to enter the physique: breathing them in, ingesting them, and skin contact **(Figure 2)**[13,39]. Among the products of airborne microplastics from urban dust that are swallowed are rubber tires and synthetic textiles [39]. Microplastics will be consumed, as was previously said, because they are widely present in the nourishment web and aquatic sources. Although the membrane's lining was too thin for MNPs to concluded, they might still get in through cuts, sweat glands, or hair follicles. While the human body has MNPs from all three pathways, the particles found in seafood and the environment pose the highest risk of exposure in absolute terms. This is brought on by the degradation of polymers over the course of time, the penetration of chemical reinforcements into the polymeric material, the existence of leftover monomers, contaminant exposure, and the replication of pathogenic microorganisms under these circumstances [40].

4.1. Exposure of the stomach

Humans inhale plastic particles mostly by ingestion, according to recent research on the toxicity and exposure of MNPs [41]. Although no studies have directly examined the noxiousness of nano plastics to people, there is indication that MPs are being obsessive over food and drink [42]. The hypothesis that people are consuming plastic particles through their food and drink is supported by the preliminary examination of human feces samples, which revealed that plastic particles were being expelled. These findings unequivocally demonstrate that humans will regularly consume microplastics and nano plastics, as Doe's research studying ingestion uptake in environmental models [43]. Nevertheless, once the MNPs elements arrive the gastrointestinal (GI) system, no research has yet looked into what happens to them. It is essential to examine how the particles move through the gastrointestinal tract and whether they cross the gut epithelium or remain in the gut lumen.

According to the available research, the essential pores located at the junction between tight channels have a maximal functional size of approximately 1.5 nm. Based on this finding, it appears improbable that MPs will infiltrate at a paracellular level [44]. That theyinfiltrate the microfold (M) cells in the Peyer's patches by endocytosis or phagocytosis is especially likely. They are also likely to penetrate through lymphatic tissue. In mice administered intraperitoneally, 1, 5, and 12 μm polymethacrylate and polystyrene elements were detected to be phagocytosed by peritoneal macrophages [45]. However, according to the data, only 0.04–0.3% of rodent models' absorption occurs through the digestive tracts.

One major concern is that nanoplastics have the ability to penetrate the intestinal epithelium and expose individuals systemically. In the past, researchers have tested polystyrene nanoparticles on a range of species both *in vivo* and *in vitro*. Ten to one hundred times more oral bioavailability of 50 nm polystyrene NPs is likely than that of microplastics (2–7%). Similar to MPs, there is no obvious association between the dimensions, structure, and consumption of NPs [46]. The amount of ingesting nanoparticles (50–500 nm) changed dramatically between different in vitro intestinal designs, according to previous studies. It varied between 1.5% and 10% depending on the kind of experimental structure utilized, as well as the size and chemical make-up of those particles.

It is important to note that samples collected from both marine and terrestrial habitats were not included in the majority of published studies, which were based on investigations utilizing models of polystyrene nanoparticles. Nevertheless, the primary polymeric resources found in these surroundings are other plastics like polypropylene (PP), polyethene (PE), and polyethene terephthalate (PET). Therefore, any inferences drawn from the results of the above-discussed research which exclusively uses polystyrene must be qualified. Instead, PP, PE, and PET trials should be a part of future model research.

4.2. Exposure to the lungs

Inhalation is the second most frequent way for humans to come into contact with nanoplastics. Airborne plastic particles from indoor surroundings, mostly from synthetic textiles, might cause occupational exposure or unintentional inhalation [47]. In outdoor settings, there is a possibility of exposure to polluted aerosols from ocean waves or airborne fertilizer particles from dry wastewater treatment plants. It is important to be mindful of this potential risk and take appropriate measures to protect oneself [48]. The lungs have a very small tissue barrier that is less than 1 μm in size and a large alveolar surface area of about 150 m2. Nanoparticles can spread throughout the entire human body because this barrier is thin enough to allow them to pass through and enter the capillary blood system [49].

The ingestion of plastic particles, especially micro and nano plastics, raises a number of detrimental health issues, including the introduction of infections and parasite vectors, chemical toxicity, and particle toxicity. This size range of particles may allow them to convert intensely lodged in the lung, where they may endure on the alveolar surface or go to added portions of the figure [50]. Inhaled plastic particles have the potential to cause lung injury. A multitude of parameters, including hydrophobicity, charge on the surface, functionalization of the surface, adjacent protein coronas, and particle size, influence the absorption and expulsion of micro and nano plastics in the lungs [51]. Furthermore, readings on animal absorption rates displays a significant association among work-related experiences and enlarged incidences of malignance and pulmonaryinflammation. Studies investigating polystyrene particle absorption rates in vitro in alveolar epithelial cells indicate that absorption is size-dependent.

According to topical investigation on the inhalation of plastic elements by individuals, urban zones' atmospheric effect is a main basis of these corpuscle [52]. Particles of synthetic fibre, comprising 29% of the total, were identified as the primary air ingredient of microplastics detected in Parisian suburban and urban regions. Based on the average yearly air flow of all fibres, fibre diameters, and fibre densities, 3–10 tons of MPs are thought to be dropped each year as a consequence of distinctive upshot. Rainfall clearly affected the observed depositions, with urban regions recording twice over the normal environment flow as suburban zones [53]. Additionally, they studied the amounts of microplastic particles in the air both indoors and outdoors attwo private apartments and one office. In the indoor samples, the fallouts indicated an absorption ranging from 1 to 60 fibers/m^3. The levels of these readings were suggestively advanced than those of the external models, which ranged from0.3 to 1.5 fibers/m^3. The majority of the enclosed illustrations were made of polypropylene, accounting for about one-third of their petrochemical origins, with the remaining samples being composed of cellulose [54]. Information about the quantity or concentration of airborne nanoplastics is currently unavailable.

4.3. Skin Contact

Nanoplastics can also be found in health and beauty products, especially in body and face washes applied topically to the skin [55]. Dermal solicitation of nanocarriers for medication delivery is another significant route of exposure. Skin penetration is significantly influenced by small particle size and stressed skin conditions, even if definitive evidence about the effects of nanocarriers is lacking [56]. The capacity of nanoplastics to pierce the skin's surface has yet to be the subject of any recent research. The possibility of textile-engineered nanoparticles piercing the epidermal barrier in minute amounts was only documented in one investigation [57]. [58] Investigated the entry of plastic particles into the physique and their consequent dispersion during the course of the skin tissue. They carried out their experiment usingpig skin tissue and fluorescent polystyrene specks with a thickness of 20–200 nm [58]. A skin-based confocal laser micrograph revealed that the hair follicles were overpopulated with 20 nm polystyrene nanoplastics that are compared to 200 nm nanoplastics. Nevertheless, neither particle was able to become embedded in the subsurface skin tissue by penetrating the stratum corneum. These results were corroborated by [58]., who found that only the outermost layers of skin may be penetrated by polystyrene particles measuring 20–200 nm in diameter [59]. Focal polystyrene nanoparticles about 40 nm in diameter were found in theperifollicular tissue of skin explants subjected to cyanoacrylate follicular stripping, according to [59]. This study determined that particles injected transcutaneously were subsequently taken up by Langerhans cells.

[60] Investigated the impact of oleic acid, ethanol, along with oleic acid-ethanol enhancers on the viatransdermal distribution of zinc oxide NPs ranging 10 nm. They found that every chemical may improve the capacity of zinc oxide nanoparticles to cross the epidermal blockade by making use of the multilamellar lipid regions that are present between corneocytes [60]. [61] developed a three-layer "sandwich model" through the examination of the crystalline structure of various lipid lamellae compositions found in stratum corneum specimens sourced from both pigs and human sources [61]. Large nanoparticles are probably prevented from entering undamaged skin by this concept.

5. POSSIBLE HARMFUL CONSEQUENCES OF NANOPLASTICS AND MICROPLASTICS ON HUMAN HEALTH:

More than a few studies implemented *in vitro* and *in vivo* have confirmed that MNPs might adversely affect oxidative stress, inflammation, necrosis, and apoptosis in addition to the immune system in humans.

5.1. Inflammation

Larger particles (202 nm and 535 nm) caused inflammatory effects on human A549 lung cells, according to an *in vitro* investigation employing different sizes of polystyrene particles. When comparing lung cells treated with bigger particles to those exposed to 64 nm particles, there was an increase in IL-8 expression [62]. The influence of 120 nm carboxylated and amino-modified polystyrene particles on the polarization of human macrophages into M1 or M2 phenotypes was studied [63]. The outcomes demonstrated that M1 markers such as CD86, NOS2, TNFα, and IL-1β continued to express themselves in the same ways. On the other hand, the introduction of both kinds of nanoparticles had an adverse effect on M2 cells' production of IL-10 and the appearance of the scavenger receptors CD163 and CD200R. High concentrations of 0.2–10 μm-sized plastic particles have been found in the Periprosthetic tissue of implants made of ultrahigh molecular weight polyethylene. Moreover, the tissues around the implant region contain macrophages, which suggests that the inflammatory response has been triggered [64].

5.2. Stress from oxidation and apoptosis

Various polystyrene nanoparticles have been demonstrated in vitro experiments to cause oxidative stress, apoptosis, and autophagic cell death in a manner that is dependent on the cell environment. It has been demonstrated, for example, that amine-modified polystyrene nanoparticles strongly bind and aggregate with mucin, thereby inducing apoptosis in intestinal epithelial cells that secrete mucin as well as non-mucin [43]. It has been shown that cationic polystyrene nanoparticles induce misfolded protein aggregation, a process that, in turn, triggers ER stress and the production of ROS generation in lung epithelial cells and mice macrophages. In BEAS-2B lung tissue cells and RAW 264.7 murine macrophages, this finally leads to autophagic cell demise. Polystyrene, whether functionalized or not, has been demonstrated to cause apoptosis in a number of human cell types, like initial human alveolar macrophages (MAC),primary human alveolar type 2 (AT2), epithelium cells, human monocytic leukaemia cell line (THP-1), individuals portrayed alveolar epithelial type 1 cell (TT1), human colon carcinoma cells (Caco-2), and human cells with lung cancer (Calu-3) [65]. Furthermore, using long non-coding RNAs (e.g., inc-61, linc-50, linc-9, and linc-2) polystyrene nanoparticles have been experimentally shown to regulate ROS in Caenorhabditis worms [66]. In mice, oral acquaintance to a amalgamation of MPs did not cause any evident severe toxicity in the liver, duodenum, ileum, jejunum, large intestine, testes, lungs, heart, spleen, or kidneys, in spite of the contaminated belongings seen in the experimental prototypes [67].

5.3. Metabolic homeostasis

Recent research has shown that microplastics and nano plastics can affect cellular metabolism in both *in vitro* and *in vivo* models, in addition to inducing inflammation and apoptosis. Because polystyrene-based nanoparticles interact with the cytoplasmic membrane, they affect the signalling pathways in airway epithelial cells.Basolateral K+ ion channels were observed to be triggered in human respiratory cells upon revelation to negative-charged carboxylated polystyrene NPs measured 20 nm [68]. By activating *ion channels* and stimulating Cl and HCO3− ion efflux, the nanoplastic particles increased short-circuit currents in a concentration-dependent manner and for an extended period of time [68]. Moreover, macrophages and the human carcinoma cell lines A549, HepG-2, and HCT116 were shown to develop sizable vesicle-like structures in response to 30 nm polystyrene nanoparticles during the endocytic pathway. Consequently, the development of binucleated cells is stimulated by blocking vesicle transit and the movement of proteins connected with cytokinesis [69]. Additionally,intestinal iron transport and cellular uptake may be interfered with by short-term oral disclosure to highly charged polystyrene NPs [70].

5.4. Immune responses

One of the main impacts linked to exposure to MPs and NPs has been found as the immune system being activated by inflammatory reactions [41]. There is little information available about how MPs and NPs affect the immune system, just like with the other toxicity factors. If one takes into account additional publications with distinct goals, such the research of wear debris from plastic prosthetic implants or those carried out using those PS nanoparticles employed as a model in pharmacological investigations, the number of immunological studies increases.

Using PS NPs for pharmacological investigations, some *in vitro* studies. observed immune cell death, reduced expression of scavenger receptors (CD163 and CD200R), and interleukin production. Using the human monocytic THP-1 cell line, [71] showed the in vitro immunological effects of fluorescent spherical PS MPs (1, 4 or 10 μm). In this instance, MPL exposure had no effect on the in vitro model of human macrophage activation or differentiation

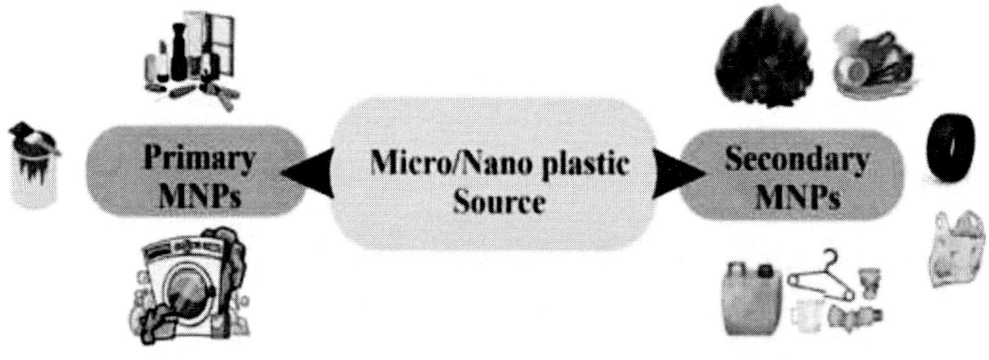

Figure 1. Major and intermediate MNP sources

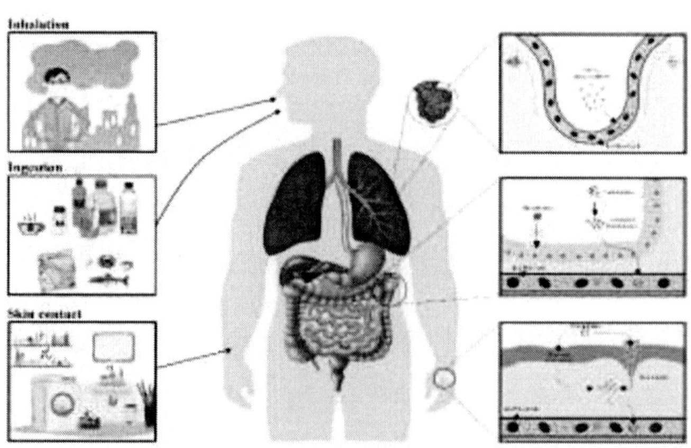

Figure 2. Entry points for plastic particles into the human body

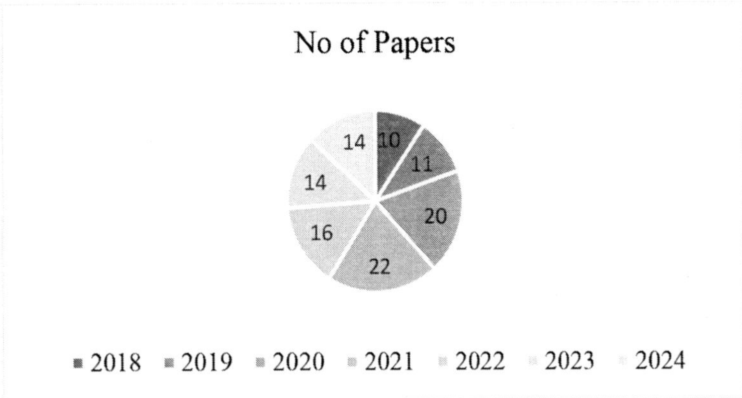

Figure 3. Distributive analysis

. Similarly, [71] when fluorescent spherical PS MPs were evaluated following 28 days of feeding, there were no indications of an inflammatory response. Furthermore, as demonstrated by elevated levels of cytokines and histamines, [72] high concentrations of amorphous PP MPs were required to stimulate the immune system, enhancing the potential hypersensitivity in human dermal fibroblast (HMC-1) cells, murine macrophages, and peripheral blood mononuclear cells. Furthermore, there was an increased expression of pro-inflammatory cytokines such histamine, TNFα, and IL-6. It's interesting to note that these effects varied in size according to concentration, with higher induction at the highest concentrations tested with particles of 20 μm in size [72].

5.5. Gut microbiome

One of the most active fields of study in recent years has been the gut microbiota, a varied collection of bacteria in the gastrointestinal tract (GIT). A complex and diverse community of microbiota lives in the gastrointestinal tract (GIT) that has developed a symbiotic relationship with its host over millennia and provides a range of benefits to it [73]. Numerous aspects of human health are impacted by the gut microbiota, and research has suggested that certain intestinal microbial populations may play a role in energy balance, which may have an impact on body weight and illnesses linked to obesity. Blood pressure control, chronic kidney disease, cardiovascular disease, and immunological state are additional health advantages [74]. In a mouse model, MP-induced intestinal dysbiosis was demonstrated [75]. Here, dysbiosis of the digestive system was brought on by changes in the gut microbiota, which ultimately disturbed physiological homeostasis. Significantly, changes in the gut microbiota have been linked to long-term illnesses of other organs, such as cancer, neurological conditions, cardiovascular conditions and renal disease [76]. The changed condition of the gut mucosa is another important factor to take into account when investigating potential negative health effects linked to MPs and NPs. Diseases associated to the intestines may serve as a modulator of internalization and its consequences. As a result, inflammatory bowel disease patients already exhibited higher particle uptake than healthy controls, which may potentially result in different unfavorable circumstances [77]. Future research should take this characteristic into account since poor diet and lifestyle choices are causing more and more changes in gut physiology in industrialized nations [78]. In connection with this, it is important to acknowledge the possible harm that MPs and NPs may cause to the gut microbiota and the ensuing consequences for human health [49].

6. CHALLENGES INVOLVED IN STUDYING THE EFFECTS OF MICROPLASTICS ON HUMANS

Because MPs and NPs are so common in our surroundings, there are a lot of challenges in planning and carrying out research to look at their negative impacts on human health. Estimating daily exposure in humans is one of the first hurdles in determining the highest exposure source since, similar to animal studies, larger concentrations are likely to be linked to negative consequences [79]. Since all research focuses on the sources of exposure, studies need to measure human exposure directly. According to a recent estimate [80], drinking water is the source that exposes people to the greatest amount of contaminants [80]. However, the analysis should have taken into account recent studies on food goods [81]. Although there are various pathways of exposure, the ones that pose the biggest hazards to human health are unknown. If we want to comprehend the consequences of MPs/NPs on human health, then developing appropriate, standardized ways to test them is a challenge and a top priority. In order to find out if MP/NPs are taken up by cells and cause alterations, we should ideally develop analytical instruments that can assess MP/NPs in human tissues de novo directly [82]. Since MPs and NPs can become entrenched in human tissue, quantifying these particles presents a number of difficulties. Sample processing, such as digestion techniques to eliminate polymers from tissues, can obliterate or change features [83]. To enable cross-national comparisons of research, techniques for regular monitoring, as well as standardized and verified sampling and analytical procedures, must be developed.

7. KNOWLEDGE GAPS

Micro- and nanoplastics (MNPLs) are widespread environmental contaminants, but their health impacts remain unclear despite years of research. Most studies have used nanopolystyrene as a model, raising concerns about the generalization of results to other microplastics. MNPLs may cause inflammatory, respiratory, intestinal, and reproductive toxicity, as shown in murine models, with increased human exposure via inhalation, food, and water in the past decade **(Table 2)**. Monitoring MNPLs in the human body is challenging due to inaccuracies in current methods, highlighting the need to improve laboratory standards, refine research methodologies, and standardize environmental quantification.

8. SUGGESTIONS AND FUTURE RECOMMENDATIONS

Addressing knowledge gaps in airborne micro- and nanoparticles (MNPs) is crucial to understanding their origin, behavior, and health impacts. Their distinct properties can trigger

9. DISTRIBUTIVE ANALYSIS

Figure 3 shows a distributive analysis of papers taken for the review process. This review analyzed Micro & Nano Plastics: Sources, fate, transport, detection and abundance in Human Health Systems, and the paper was taken from reputed sources like Google Scholar, Elsevier, Springer, etc., in the year 2018; we have taken 10 papers. In 2019, 11 papers were analyzed. 20 papers were reviewed in 2020. In the year 2021, 22 papers will be reviewed. In 2022, 16 papers were taken. In 2023, 14 papers were analyzed, and in the year 2024, 14 papers were reviewed.

10. CONCLUSION

Global emphasis has been focused on micro and nano plastics (MNPs) because of their pervasiveness in the environment. The health of people is seriously threatened by these tiny plastic particles, which are found in soil, water, air, and the biosphere. The data regarding the potential effects of $MPs\ and\ NPs$ on human health was reviewed in this review. One major obstacle to a thorough evaluation of MNPs' effects on human health is the need for more knowledge regarding their sources, occurrences, behaviours, and health hazards in the human body despite their widespread prevalence. In order to gain a deeper comprehension of the impact of MNPs on the human body, it is critical to recognize the current knowledge gaps that need to be filled right away and offer viable answers. The greatest obstacles to current reviews on MNPs in the human body, according to our examination of 107 articles from a variety of sources, including Google Scholar, Springer, and Elsevier, are the complexity of MNPs and the requirement for effective analytical techniques. As a result, the creation of a uniform and standard analytical technique is required. Additionally, we noted important gaps in our understanding that need to be filled in order to evaluate the impact of MNPs on human healthfully. Humans are exposed to MNPs on a daily basis, so further research is necessary to understand better any possible harm they may cause to human health.

Table 1. Sources of microplastics and nanoplastics in the environment

MNPs Source	Properties	Environmental Entry point	Application	References
Microbead personal care products	Microbead varying in color and ~100-1000 µm	Sewage Sludge	Film-foaming, Exfoliating in personal care products	84
Paint	Between 0.3 and 5mm from synthetic polymers mainly epoxy resins and alkyds	Surface runoff	Marine coating, road-making paint	85
Textile Fabric	100-1000µm MNPs from Nylon Fabric and acrylic	Wastewater	Functionality in synthetic fabric	86
Fishing Gears	Polyamide and polypropelene monofilaments	Loss/Drifting	N/A	87
Construction Industry	Polyurethane Polymers, polyethylene, polyvinylchloride	Surface runoff/Drifting	N/A	88
Municipal debris	Plastic bottles, Fragments of Plastic bags and other packing materials	Surface runoff	N/A	89
Plastic pellets	Generally with a diameter of 2-5 mm and typical shape	Drifting	Building blocks for nearly each products by plastics	90
Sewage treatment effluents	Various kinds of MNPs Industrial production of plastics, chemical laundry products	N/A	N/A	91

Table 2. Exposure to nanoparticles and microplastics and their effects on specific organs

Health Effects	Potential diseases	Comments	References
Neurological effects	Effects of Neurotoxic (dizziness, headache, fatigue)	correlation with exposure to styrene in the environment; further study is required for definitive proof	92, 93.
Reproductive	Reduced offspring weight, gonadal damage, decreased fertility	Research on human implications still needs to be completed; however, studies on animals have shown effects.	94,95.
Respiratory system	Respiratory issues like wheezing, coughing, dyspnea and lung cancer	The research was conducted primarily in work environments, and some illnesses were connected to particular substances like polyacrylic NPs and styrene.	96,97.
GI Tract	63 gut barrier dysfunction, 60 colorectal cancer, IBD, 62 pancreatic cancer	Fecal MPs and IBS severity positively correlate; human studies on the impact of the microbiome are lacking	98.
Skin	Allergic infection	few human research, primarily focusing on allergic reactions	99, 100.
Blood neoplasms	Bladder cancer, Leukaemia	associations discovered with prolonged exposure in work environments	101,102.
Osteolysis around bone prosthesis	Periprosthetic osteolysis	connected to PE particles found in hip replacements	103, 104.

ACKNOWLEDGMENTS

We acknowledge the management of the Vellore Institute of Technology (VIT), Vellore for providing lab facilities to carry out the research work.

LIST OF ABBREVIATIONS

MNPs - Micro and Nano plastics

MPs - Micro plastic

NPs - Nano plastic

PE - Polyethylene

PLA - Polylactic acid

PCL - Polycaprolactone

PP - Polypropylene

PET - Polyethylene terephthalate

PC - Polycarbonate

PA - Polyamide

REFERENCES

1. Yuan, Z., Nag, R., & Cummins, E. (2022). Human health concerns regarding microplastics in the aquatic environment-From marine to food systems. *Science of the Total Environment*, *823*, 153730.
2. Birch, Q. T., Potter, P. M., Pinto, P. X., Dionysiou, D. D., & Al-Abed, S. R. (2020). Sources, transport, measurement and impact of nano and microplastics in urban watersheds. *Reviews in Environmental Science and Bio/Technology*, *19*, 275-336.
3. Deji-Oloruntoba, O., Agidigbi, T. S., & Jang, M. (2024). Microplastics and Nano-plastics Contamination in Foods: Current Understanding of the Health Impact on Human and Potential Solutions. *European Journal of Nutrition & Food Safety*, *16*(5), 11-31.
4. Chen, C. Y., & Lin, Z. (2024). Exploring the potential and challenges of developing physiologically-based toxicokinetic models to support human health risk assessment of microplastic and nanoplastic particles. *Environment International*, 108617.
5. El Hadri, H., Giechaskiel, B., Al Wasif Ruiz, T., Putaud, J. P., Geiss, O., & Barrero, J. (2024). Micro-and nanoplastics in the air: Sources, transport & analytical methods for their determination.
6. Dey, S., Veerendra, G. T. N., Babu, P. A., Manoj, A. P., & Nagarjuna, K. (2024). Degradation of plastics waste and its effects on biological ecosystems: A scientific analysis and comprehensive review. *Biomedical Materials & Devices*, *2*(1), 70-112.
7. Lopez-Lorenzo, X. (2024). *Chemoenzymatic Synthesis and Degradation of Plastics* (Doctoral dissertation, KTH Royal Institute of Technology).
8. Lima, V., Hossain, U. H., Walbert, T., Seidl, T., & Ensinger, W. J. R. P. (2018). Mass spectrometric comparison of swift heavy ion-induced and anaerobic thermal degradation of polymers. *Radiation Physics and Chemistry*, *144*, 21-28.
9. Mayekar, P. C., Castro-Aguirre, E., Auras, R., Selke, S., & Narayan, R. (2020). Effect of nano-clay and surfactant on the biodegradation of poly (lactic acid) films. *Polymers*, *12*(2), 311.
10. Di Giulio, T., De Benedetto, G. E., Ditaranto, N., Malitesta, C., & Mazzotta, E. (2024).Insights into Plastic Degradation Processes in Marine Environment by X-ray Photoelectron Spectroscopy Study. *International Journal of Molecular Sciences*, *25*(10), 5060.
11. Mandal, M., Roy, A., Popek, R., & Sarkar, A. (2024). Micro-and Nano-Plastic Degradation by Bacterial Enzymes: A Solution to 'White Pollution'. *The Microbe*, 100072.
12. Mohamed Nor, N. H., Kooi, M., Diepens, N. J., & Koelmans, A. A. (2021). Lifetime accumulation of microplastic in children and adults. *Environmental science & technology*, *55*(8), 5084-5096.
13. Prata, J. C., da Costa, J. P., Lopes, I., Duarte, A. C., & Rocha-Santos, T. (2020). Environmental exposure to microplastics: An overview on possible human health effects. *Science of the total environment*, *702*, 134455.
14. Molina, E., & Benedé, S. (2022). Is there evidence of health risks from exposure to micro-and nanoplastics in foods?. *Frontiers in Nutrition*, *9*, 910094.
15. Udovicki, B., Andjelkovic, M., Cirkovic-Velickovic, T., & Rajkovic, A. (2022). Microplastics in food: scoping review on health effects, occurrence, and human exposure. *International Journal of Food Contamination*, *9*(1), 7.
16. Eerkes-Medrano, D., Leslie, H. A., & Quinn, B. (2019). Microplastics in drinking water: A review and assessment. *Current Opinion in Environmental Science & Health*, *7*, 69-75.
17. Evangeliou, N., Grythe, H., Klimont, Z., Heyes, C., Eckhardt, S., Lopez-Aparicio, S., & Stohl, A. (2020). Atmospheric transport is a major pathway of microplastics to remote regions. *Nature communications*, *11*(1), 3381.
18. Gonçalves, J. M., & Bebianno, M. J. (2021). Nanoplastics impact on marine biota: A review. *Environmental Pollution*, *273*, 116426.
19. da Costa, J. P. (2018). Micro-and nanoplastics in the environment: research and policymaking. *Current Opinion in Environmental Science & Health*, *1*, 12-16.
20. Ma, H., Pu, S., Liu, S., Bai, Y., Mandal, S., & Xing, B. (2020). Microplastics in aquatic environments: toxicity to trigger ecological consequences. *Environmental Pollution*, *261*, 114089.
21. Smyth, K., Drake, J., Li, Y., Rochman, C., Van Seters, T., & Passeport, E. (2021). Bioretention cells remove microplastics from urban stormwater. *Water research*, *191*, 116785.

22. Guerranti, C., Martellini, T., Perra, G., Scopetani, C., & Cincinelli, A. (2019). Microplastics in cosmetics: Environmental issues and needs for global bans. *Environmental toxicology and pharmacology*, *68*, 75-79.
23. Dalili, A., Samiei, E., & Hoorfar, M. (2019). A review of sorting, separation and isolation of cells and microbeads for biomedical applications: microfluidic approaches. *Analyst*, *144*(1), 87-113.
24. Alves, T. F., Morsink, M., Batain, F., Chaud, M. V., Almeida, T., Fernandes, D. A., ...& Severino, P. (2020). Applications of natural, semi-synthetic, and synthetic polymers in cosmetic formulations. *Cosmetics*, *7*(4), 75.
25.Ding, N., An, D., Yin, X., & Sun, Y. (2020). Detection and evaluation of microbeads and other microplastics in wastewater treatment plant samples. *Environmental Science and Pollution Research*, *27*, 15878-15887.
26.Karbalaei, S., Hanachi, P., Walker, T. R., & Cole, M. (2018). Occurrence, sources, human health impacts and mitigation of microplastic pollution. *Environmental science and pollution research*, *25*, 36046-36063.
27. An, L., Liu, Q., Deng, Y., Wu, W., Gao, Y., & Ling, W. (2020). Sources of microplastic in the environment. *Microplastics in terrestrial environments: Emerging contaminants and major challenges*, 143-159.
28. Kitahara, K. I., & Nakata, H. (2020). Plastic additives as tracers of microplastic sources in Japanese road dusts. *Science of the Total Environment*, *736*, 139694.
29. Galvão, A., Aleixo, M., De Pablo, H., Lopes, C., & Raimundo, J. (2020). Microplastics in wastewater: microfiber emissions from common household laundry. *Environmental Science and Pollution Research*, *27*, 26643-26649.
30. Battulga, B., Kawahigashi, M., & Oyuntsetseg, B. (2019). Distribution and composition of plastic debris along the river shore in the Selenga River basin in Mongolia. *Environmental Science and Pollution Research*, *26*, 14059-14072.
31. Wang, J., Zhao, X., Wu, F., Niu, L., Tang, Z., Liang, W., ...& Wang, X. (2021). Characterization, occurrence, environmental behaviors, and risks of nanoplastics in the aquatic environment: current status and future perspectives. *Fundamental Research*, *1*(3), 317-328.
32. Liu, Y., Shao, H., Liu, J., Cao, R., Shang, E., Liu, S., & Li, Y. (2021). Transport and transformation of microplastics and nanoplastics in the soil environment: A critical review. *Soil use and management*, *37*(2), 224-242.
33. Kiran, B. R., Kopperi, H., & Venkata Mohan, S. (2022). Micro/nano-plastics occurrence, identification, risk analysis and mitigation: challenges and perspectives. *Reviews in environmental science and bio/technology*, *21*(1), 169-203.
34. Okeke, E. S., Chukwudozie, K. I., Addey, C. I., Okoro, J. O., Ezeorba, T. P. C., Atakpa, E. O., ... & Nwuche, C. O. (2023). Micro and nanoplastics ravaging our agroecosystem: A review of occurrence, fate, ecological impacts, detection, remediation, and prospects. *Heliyon*, *9*(2).
35. Hurley, R. R., & Nizzetto, L. (2018). Fate and occurrence of micro (nano) plastics in soils: Knowledge gaps and possible risks. *Current Opinion in Environmental Science & Health*, *1*, 6-11.
36. Zhou, Y., Wang, J., Zou, M., Jia, Z., Zhou, S., & Li, Y. (2020). Microplastics in soils: A review of methods, occurrence, fate, transport, ecological and environmental risks. *Science of the Total Environment*, *748*, 141368.
37. Vighi, M., Bayo, J., Fernández-Piñas, F., Gago, J., Gómez, M., Hernández-Borges, J., ...& Rosal, R. (2021). Micro and nano-plastics in the environment: Research priorities for the near future. *Reviews of Environmental Contamination and Toxicology Volume 257*, 163-218.
38. Mendoza, L. M. R., Vargas, D. L., & Balcer, M. (2021). Microplastics occurrence and fate in the environment. *Current Opinion in Green and Sustainable Chemistry*, *32*, 100523.
39. Rahman, A., Sarkar, A., Yadav, O. P., Achari, G., & Slobodnik, J. (2021). Potential human health risks due to environmental exposure to nano-and microplastics and knowledge gaps: A scoping review. *Science of the Total Environment*, *757*, 143872.
40. Camacho, M., Herrera, A., Gómez, M., Acosta-Dacal, A., Martínez, I., Henríquez-Hernández, L. A., & Luzardo, O. P. (2019). Organic pollutants in marine plastic debris from Canary Islands beaches. *Science of the Total Environment*, *662*, 22-31.
41. Lehner, R., Weder, C., Petri-Fink, A., & Rothen-Rutishauser, B. (2019).Emergence of nanoplastic in the environment and possible impact on human health. *Environmental science & technology*, *53*(4), 1748-1765.
42. Mason, S. A., Welch, V. G., & Neratko, J. (2018). Synthetic polymer contamination in bottled water. *Frontiers in chemistry*, *6*, 389699.
43. Winiarska, E., Jutel, M., & Zemelka-Wiacek, M. (2024). The potential impact of nano-and microplastics on human health: Understanding human health risks. *Environmental Research*, 118535.

44. Kurtuluş, E. M., & Çöl, B. G. (2023). Micro to nano plastics and its link to human health. *International Journal of Medical Biochemistry*, *6*(1), 51.
45. González-Acedo, A., García-Recio, E., Illescas-Montes, R., Ramos-Torrecillas, J., Melguizo-Rodríguez, L., & Costela-Ruiz, V. J. (2021). Evidence from in vitro and in vivo studies on the potential health repercussions of micro-and nanoplastics. *Chemosphere*, *280*, 130826.
46. Stapleton, P. A. (2019). *Toxicological considerations of nano-sized plastics. AIMS Environ. Sci. 6, 367–378.*
47. Rist, S., Almroth, B. C., Hartmann, N. B., & Karlsson, T. M. (2018). A critical perspective on early communications concerning human health aspects of microplastics. *Science of the Total Environment*, *626*, 720-726.
48. Facciolà, A., Visalli, G., Pruiti Ciarello, M., & Di Pietro, A. (2021). Newly emerging airborne pollutants: current knowledge of health impact of micro and nanoplastics. *International journal of environmental research and public health*, *18*(6), 2997.
49. Rodrigues, A. C. B., de Jesus, G. P., Waked, D., Gomes, G. L., Silva, T. M., Yariwake, V. Y., ... & Veras, M. M. (2022). Scientific evidence about the risks of micro and nanoplastics (MNPLs) to human health and their exposure routes through the environment. *Toxics*, *10*(6), 308.
50. Kumar, R., Manna, C., Padha, S., Verma, A., Sharma, P., Dhar, A., ...& Bhattacharya, P. (2022). Micro (nano) plastics pollution and human health: How plastics can induce carcinogenesis to humans?. *Chemosphere*, *298*, 134267.
51. Bastyans, S., Jackson, S., & Fejer, G. (2022). Micro and nano-plastics, a threat to human health?. *Emerging topics in life sciences*, *6*(4), 411-422.
52. Yang, S., Cheng, Y., Chen, Z., Liu, T., Yin, L., Pu, Y., & Liang, G. (2021). In vitro evaluation of nanoplastics using human lung epithelial cells, microarray analysis and co-culture model. *Ecotoxicology and environmental safety*, *226*, 112837.
53. Alimba, C. G., Faggio, C., Sivanesan, S., Ogunkanmi, A. L., & Krishnamurthi, K. (2021). Micro (nano)-plastics in the environment and risk of carcinogenesis: Insight into possible mechanisms. *Journal of Hazardous Materials*, *416*, 126143.
54. Cary, C. M., Seymore, T. N., Singh, D., Vayas, K. N., Goedken, M. J., Adams, S., ...& Stapleton, P. A. (2023). Single inhalation exposure to polyamide micro and nanoplastic particles impairs vascular dilation without generating pulmonary inflammation in virgin female Sprague Dawley rats. *Particle and Fibre Toxicology*, *20*(1), 16.
55. Mariano, S., Tacconi, S., Fidaleo, M., Rossi, M., & Dini, L. (2021). Micro and nanoplastics identification: classic methods and innovative detection techniques. *Frontiers in toxicology*, *3*, 636640.
56. Biswas, S., Bagchi, D., & Ghosh, D. (2022). The effects of (micro and nano) plastics on the human body: nervous system, respiratory system, digestive system, placental barrier, skin, and excretory system. In *Assessing the Effects of Emerging Plastics on the Environment and Public Health* (pp. 148-171).IGI Global.
57. Cheng, Y., Yang, S., Yin, L., Pu, Y., & Liang, G. (2023). Recent consequences of micro-nanaoplastics (MNPLs) in subcellular/molecular environmental pollution toxicity on human and animals. *Ecotoxicology and environmental safety*, *249*, 114385.
58. Feng, Y., Tu, C., Li, R., Wu, D., Yang, J., Xia, Y., ...& Luo, Y. (2023). A systematic review of the impacts of exposure to micro-and nano-plastics on human tissue accumulation and health. *Eco-Environment & Health*.
59. Schmidt, A., da Silva Brito, W. A., Singer, D., Mühl, M., Berner, J., Saadati, F., ...& Bekeschus, S. (2023). Short-and long-term polystyrene nano-and microplastic exposure promotes oxidative stress and divergently affects skin cell architecture and Wnt/beta-catenin signaling. *Particle and fibre toxicology*, *20*(1), 3.
60. Ramsperger, A. F., Bergamaschi, E., Panizzolo, M., Fenoglio, I., Barbero, F., Peters, R., ...& Laforsch, C. (2023). Nano-and microplastics: a comprehensive review on their exposure routes, translocation, and fate in humans. *NanoImpact*, *29*, 100441.
61. Domenech, J., Annangi, B., Marcos, R., Hernández, A., & Catalán, J. (2023). Insights into the potential carcinogenicity of micro-and nano-plastics. *Mutation Research/Reviews in Mutation Research*, *791*, 108453.
62. Yee, M. S. L., Hii, L. W., Looi, C. K., Lim, W. M., Wong, S. F., Kok, Y. Y., ... & Leong, C. O. (2021). Impact of microplastics and nanoplastics on human health. *Nanomaterials*, *11*(2), 496.
63. Khan, A., & Jia, Z. (2023). Recent insights into uptake, toxicity, and molecular targets of microplastics and nanoplastics relevant to human health impacts. *Iscience*, *26*(2).
64. Rubio, L., Marcos, R., & Hernández, A. (2020). Potential adverse health effects of ingested micro-and nanoplastics on humans. Lessons learned from in vivo and in vitro mammalian models. *Journal of Toxicology and Environmental Health, Part B*, *23*(2), 51-68.

65. Shi, Q., Tang, J., Liu, R., & Wang, L. (2022). Toxicity in vitro reveals potential impacts of microplastics and nanoplastics on human health: A review. *Critical Reviews in Environmental Science and Technology*, *52*(21), 3863-3895.
66. Chang, X., Xue, Y., Li, J., Zou, L., & Tang, M. (2020). Potential health impact of environmental micro-and nanoplastics pollution. *Journal of Applied Toxicology*, *40*(1), 4-15.
67. Weber, A., Schwiebs, A., Solhaug, H., Stenvik, J., Nilsen, A. M., Wagner, M., ...& Radeke, H. H. (2022). Nanoplastics affect the inflammatory cytokine release by primary human monocytes and dendritic cells. *Environment international*, *163*, 107173.
68. Zhu, X., Wang, C., Duan, X., Liang, B., Xu, E. G., & Huang, Z. (2023). Micro-and nanoplastics: a new cardiovascular risk factor?. *Environment international*, *171*, 107662.
69. Yin, J., Ju, Y., Qian, H., Wang, J., Miao, X., & Zhu, Y. *Nanoplastics and microplastics may be damaging our livers. Toxics. 2022; 10 (10): 586.*
70. Gopinath, P. M., Saranya, V., Vijayakumar, S., Mythili Meera, M., Ruprekha, S., Kunal, R., ...& Chandrasekaran, N. (2019). Assessment on interactive prospectives of nanoplastics with plasma proteins and the toxicological impacts of virgin, coronated and environmentally released-nanoplastics. *Scientific Reports*, *9*(1), 8860.
71. Stock, V., Böhmert, L., Lisicki, E., Block, R., Cara-Carmona, J., Pack, L. K., ...& Lampen, A. (2019). Uptake and effects of orally ingested polystyrene microplastic particles in vitro and in vivo. *Archives of toxicology*, *93*, 1817-1833.
72. Hwang, J., Choi, D., Han, S., Choi, J., & Hong, J. (2019). An assessment of the toxicity of polypropylene microplastics in human derived cells. *Science of the Total Environment*, *684*, 657-669.
73. Qiao, J., Chen, R., Wang, M., Bai, R., Cui, X., Liu, Y., ...& Chen, C. (2021). Perturbation of gut microbiota plays an important role in micro/nanoplastics-induced gut barrier dysfunction. *Nanoscale*, *13*(19), 8806-8816.
74. Ali, N., Katsouli, J., Marczylo, E. L., Gant, T. W., Wright, S., & de la Serna, J. B. (2024). The potential impacts of micro-and-nano plastics on various organ systems in humans. *EBioMedicine*, *99*.
75. Remmelts, M. (2021). *Effect of nano-and microplastics on the human immune system and their influence on inflammatory bowel disease* (Doctoral dissertation).
76. Covello, C., Di Vincenzo, F., Cammarota, G., & Pizzoferrato, M. (2024). Micro (nano) plastics and their potential impact on human gut health: a narrative review. *Current Issues in Molecular Biology*, *46*(3), 2658-2677.
77. Niu, H., Liu, S., Jiang, Y., Hu, Y., Li, Y., He, L., ...& Lou, X. (2023). Are microplastics toxic? A review from eco-toxicity to effects on the gut microbiota. *Metabolites*, *13*(6), 739.
78. Adler, M. Y., Issoual, I., Rückert, M., Deloch, L., Meier, C., Tschernig, T., ...& Paulsen, F. (2024). Effect of micro-and nanoplastic particles on human macrophages. *Journal of Hazardous Materials*, *471*, 134253.
79. Jin, Y., Xia, J., Pan, Z., Yang, J., Wang, W., & Fu, Z. (2018). Polystyrene microplastics induce microbiota dysbiosis and inflammation in the gut of adult zebrafish. *Environmental Pollution*, *235*, 322-329.
80. Senathirajah, K., Attwood, S., Bhagwat, G., Carbery, M., Wilson, S., & Palanisami, T. (2021). Estimation of the mass of microplastics ingested–A pivotal first step towards human health risk assessment. *Journal of Hazardous Materials*, *404*, 124004.
81. Conti, G. O., Ferrante, M., Banni, M., Favara, C., Nicolosi, I., Cristaldi, A., ...& Zuccarello, P. (2020). Micro-and nano-plastics in edible fruit and vegetables. The first diet risks assessment for the general population. *Environmental Research*, *187*, 109677.
82. Blackburn, K., & Green, D. (2022). The potential effects of microplastics on human health: What is known and what is unknown. *Ambio*, *51*(3), 518-530.
83. Thornton Hampton, L. M., Bouwmeester, H., Brander, S. M., Coffin, S., Cole, M., Hermabessiere, L., ...& Weisberg, S. B. (2022). Research recommendations to better understand the potential health impacts of microplastics to humans and aquatic ecosystems. *Microplastics and Nanoplastics*, *2*(1), 18.
84. Nel, H. A., Dalu, T., Wasserman, R. J., & Hean, J. W. (2019). Colour and size influences plastic microbead underestimation, regardless of sediment grain size. *Science of the Total Environment*, *655*, 567-570.
85. Gaylarde, C. C., Neto, J. A. B., & da Fonseca, E. M. (2021). Paint fragments as polluting microplastics: A brief review. *Marine Pollution Bulletin*, *162*, 111847.
86. Cai, Y., Yang, T., Mitrano, D. M., Heuberger, M., Hufenus, R., & Nowack, B. (2020). Systematic study of microplastic fiber release from 12 different polyester textiles during washing. *Environmental Science & Technology*, *54*(8), 4847-4855.
87. Xue, B., Zhang, L., Li, R., Wang, Y., Guo, J., Yu, K., & Wang, S. (2020). Underestimated microplastic pollution derived from fishery activities and "hidden" in deep sediment. *Environmental science & technology*, *54*(4), 2210-2217.

88. Xu, C., Zhang, B., Gu, C., Shen, C., Yin, S., Aamir, M., & Li, F. (2020). Are we underestimating the sources of microplastic pollution in terrestrial environment?. *Journal of hazardous materials*, *400*, 123228.
89. Sobhani, Z., Lei, Y., Tang, Y., Wu, L., Zhang, X., Naidu, R., ...& Fang, C. (2020). Microplastics generated when opening plastic packaging. *Scientific reports*, *10*(1), 4841.
90. Mendoza, L. M. R., Karapanagioti, H., & Álvarez, N. R. (2018). Micro (nanoplastics) in the marine environment: current knowledge and gaps. *Current Opinion in Environmental Science & Health*, *1*, 47-51.
91. Mak, C. W., Tsang, Y. Y., Leung, M. M. L., Fang, J. K. H., & Chan, K. M. (2020). Microplastics from effluents of sewage treatment works and stormwater discharging into the Victoria Harbor, Hong Kong. *Marine Pollution Bulletin*, *157*, 111181.
92. Yan, Z., Liu, Y., Zhang, T., Zhang, F., Ren, H., & Zhang, Y. (2021). Analysis of microplastics in human feces reveals a correlation between fecal microplastics and inflammatory bowel disease status. *Environmental science & technology*, *56*(1), 414-421.
93. Liu, S., He, Y., Yin, J., Zhu, Q., Liao, C., & Jiang, G. (2024). Neurotoxicities induced by micro/nanoplastics: A review focusing on the risks of neurological diseases. *Journal of Hazardous Materials*, 134054.
94. Chen, X., Zhuang, J., Chen, Q., Xu, L., Yue, X., & Qiao, D. (2022). Polyvinyl chloride microplastics induced gut barrier dysfunction, microbiota dysbiosis and metabolism disorder in adult mice. *Ecotoxicology and Environmental Safety*, *241*, 113809.
95. Sun, L., Liao, K., & Wang, D. (2021). Comparison of transgenerational reproductive toxicity induced by pristine and amino modified nanoplastics in Caenorhabditis elegans. *Science of the Total Environment*, *768*, 144362.
96. Sartorelli, P., D'Hauw, G. A. B. R. I. E. L. E., Spina, D., Volterrani, L., & Mazzei, M. A. (2020). A case of hypersensitivity pneumonitis in a worker exposed to terephthalic acid in the production of polyethylene terephthalate. *International Journal of Occupational Medicine and Environmental Health*, *33*(1), 119-123.
97. Huang, X., Saha, S. C., Saha, G., Francis, I., & Luo, Z. (2024). Transport and deposition of microplastics and nanoplastics in the human respiratory tract. *Environmental Advances*, *16*, 100525.
98. Auguet, T., Bertran, L., Barrientos-Riosalido, A., Fabregat, B., Villar, B., Aguilar, C., & Sabench, F. (2022). Are ingested or inhaled microplastics involved in nonalcoholic fatty liver disease?. *International Journal of Environmental Research and Public Health*, *19*(20), 13495.
99. Aristizabal, M., Jiménez-Orrego, K. V., Caicedo-León, M. D., Páez-Cárdenas, L. S., Castellanos-García, I., Villalba-Moreno, D. L., ... & Gold, M. (2024). Microplastics in dermatology: Potential effects on skin homeostasis. *Journal of Cosmetic Dermatology*, *23*(3), 766-772.
100. Pecorelli, A., & Valacchi, G. Microplastics and Nanoplastics Contamination: An Emerging Environmental Issue for Skin Health. In *Environmental Stressors and OxInflammatory Tissues Responses* (pp. 136-152). CRC Press.
101. Werder, E. J., Engel, L. S., Richardson, D. B., Emch, M. E., Gerr, F. E., Kwok, R. K., & Sandler, D. P. (2018). Environmental styrene exposure and neurologic symptoms in US Gulf coast residents. *Environment international*, *121*, 480-490.
102. Rajendran, D., & Chandrasekaran, N. (2023). Journey of micronanoplastics with blood components. *RSC advances*, *13*(45), 31435-31459.
103. Yu, C. W., Luk, T. C., & Liao, V. H. C. (2021). Long-term nanoplastics exposure results in multi and trans-generational reproduction decline associated with germline toxicity and epigenetic regulation in Caenorhabditis elegans. *Journal of hazardous materials*, *412*, 125173.
104. Sun, A., & Wang, W. X. (2023). Human exposure to microplastics and its associated health risks. *Environment & Health*, *1*(3), 139-149.

ENGINEERED LIPOSOMAL NANOCARRIERS FOR TARGETED CHEMOTHERAPEUTIC DELIVERY IN LUNG ADENOCARCINOMA

Mahalakshmi Nannan[1], Sivaramakrishnan Venkatabalasubramanian[1*]

[1]Department of Genetic Engineering, Faculty of Engineering and Technology, SRM Institute of Science and Technology, Kattankulathur, Chennai – 603205, Tamil Nadu, India.
Corresponding author: Sivaramakrishnan Venkatabalasubramanian,
Corresponding author mail ID: sivaramv@srmist.edu.in

ABSTRACT

This study aimed to deliver an innovative drug delivery system through liposome nanocarrier by thin film hydration method. The outer membranes of liposomes are modified through the surface attachment with the PEG molecule or any other targeting agent to enhance its specificity. We will discuss the formulation and characterization of chemotherapeutic drug-loaded PEGylated liposomes synthesized against lung adenocarcinoma. Dasatinib (former BMS 354825) is an orally available small-molecule multikinase inhibitor, was entrapped inside the small, lamellar liposomes composed of Lecithin, cholesterol, PEG (MPEG-DSPE2000), and DOPC. These liposomes were characterized by various techniques such as ATR-FTIR, to study the interaction between drug and excipient while the outer morphology was characterized through SEM, and TEM (microscopy techniques). TEM and SEM presented that formulated liposomes were spherical-shaped with a smoothened surface in the periphery. Particle size, PDI, Zeta Potential & Entrapment efficiency were calculated. The in-vitro drug release study of all the formulations was carried out at 0,2,4,6,12 and 48 h, which optimized formulation release in 24 h. The stability study of the optimized formulation was performed and found that the suitable storage condition there is 4° C. The present study focuses on enhancing these characteristics of the solitary drug, leading to the design of liposomal formulation for the passive treatment of lung adenocarcinoma.

KEYWORDS: *Chemotherapeutic drug, Liposome, Thin film technique, Nanoparticles, Lung Adenocarcinoma.*

INNOVATIVE HYDROGEL THERAPY: UTILIZING CHICKEN FEATHER KERATINAND PLANT EXTRACT FOR ICHTHYOSIS VULGARIS

Swalakshana A.S

Final year Student, Department of biotechnology, Sri Venkateswara College Of Engineering, Sriperumbudur, Pennalur Village, Chennai - Bengaluru High Road, Sriperumbudur Tk. – 602117.

Author mail id: 2021bt0408@svce.ac.in

ABSTRACT

Ichthyosis vulgaris, affecting approximately 1 in 250 individuals globally is a hereditary skin disorder that occurs due to the mutation of the filaggrin gene which is a protein essential for keratin aggregation, skin barrier formation and acts as a natural moisturizing agent in the skin. It is characterized by dry, scaly skin due to the abnormal shedding of skin cells, leading to discomfort and affects the quality of life of those affected. Our project aims to treat this condition using a hydrogel prepared from chicken feather keratin infused with *Glycyrrhiza glabra* root extract. Chicken feather which is a primary waste produced in the poultry industry, has a high percentage of keratin when compared to other sources. This keratin is a structural protein that has the potential of skin regeneration and repair. In our work, the extraction of chicken feather was carried using sodium sulfide and HCl method as it efficiently breaks keratin bonds and result in high quality keratin extraction. The extract was purified and the characterization studies were performed to confirm its potential application in treating Ichthyosis vulgaris. *Glycyrrhiza glabra* is plant has compounds such as glycyrrhizin, flavonoids, Licochalcone A and Glabridin which together has properties such as anti-inflammatory, anti-oxidant and hydrating agent. These compounds help in reducing hyper pigmentation caused by Ichthyosis vulgaris and aids in cell renewal, prevent skin barrier damage and promote skin repair and regeneration. The hydrogel efficiency is evaluated by performing, RT-PCR and qPCR to analyze filaggrin gene expression pre- and post-treatment. Additionally, cell adhesion and cell viability assays will be conducted to assess the hydrogel's impact on skin cell health and functionality. Keratin promotes skin repair and regeneration, while the plant extract, rich in glycyrrhizin and flavonoids, provides anti-inflammatory, antioxidant, and hydrating benefits. This formulation addresses key symptoms like dryness, scaling, and hyperpigmentation, offering an effective and eco-friendly solution for managing the condition.

KEYWORDS: *Chicken feather keratin, Glycyrrhiza glabra, Hydrogel, Ichthyosis Vulgaris, Skin regeneration, Gene expression, Cell division.*

IDENTIFICATION AND MOLECULAR CHARACTERIZATION OF FELINE CALICIVIRUS (FCV) FROM SUSPECTED CATS

Sridevi R[1], Raja P[2], Vijayalakshmi R[2], Vinitha V[2], Parthiban M[2]

[1]*Department of Biotechnology, Rajalakshmi Engineering College, Chennai*
[2]*Department of Animal Biotechnology, Madras Veterinary College, Chennai*
Author mail id: sridevi2001ragu@gmail.com

ABSTRACT

Feline calicivirus (FCV) is a highly contagious viral pathogen posing a significant threat to feline populations globally. This study aimed to detect and molecularly characterize FCV in clinically suspected cats from Chennai, Tamil Nadu. A total of 50 oral swab samples were collected from cats presenting clinical signs consistent with FCV infection between November 2024 and January 2025. Total RNA was extracted and analysed via reverse transcription-polymerase chain reaction (RT-PCR) targeting the VP1 gene, a critical region for FCV detection and characterization. Among the samples, 22 (44%) tested positive for FCV, confirming the virus's substantial prevalence in the region. To elucidate the genetic diversity of the circulating FCV strains, complete VP1 gene amplification and sequencing were performed on the positive samples. Phylogenetic analysis demonstrated close genetic relatedness of the detected strains to previously reported FCV isolates from India and other regions. Notably, unique nucleotide and amino acid substitutions were identified, underscoring the genetic variability within the FCV population. The findings highlight the significant prevalence and molecular diversity of FCV in Chennai, with implications for disease monitoring and management. The observed genetic variations within the VP1 gene emphasize the need for ongoing molecular surveillance to track the evolution of FCV strains. These insights are vital for the development and optimization of effective vaccination strategies tailored to regional virus strains, contributing to improved control and prevention of FCV infections in feline populations.

KEYWORDS: *Feline Calicivirus, RT-PCR, Phylogenetics, Capsid protein, Genetic diversity.*

IDENTIFICATION AND GENETIC CHARACTERIZATION OF FELINE PANLEUKOPENIA VIRUS: A STUDY ON CLINICALLY SUSPECTED CATS

Niroshini S[1], Raja P[2], R.Vijayalakshmi[2], V.Vinitha[2], M.Parthiban[2]

[1]Department of Biotechnology, Rajalakshmi Engineering College, Chennai
[2]Department of Animal Biotechnology, Madras Veterinary College, Chennai
Author mail id: sniroshini02@gmail.com

ABSTRACT

Feline panleukopenia virus (FPV) is a highly contagious and lethal virus affecting cats, poses a significant threat to feline populations worldwide. The present study was aimed to detect and characterization of FPV from clinically suspected cases of feline panleukopenia by molecular methods. A total of 41 faecal samples from suspected FPV cases were collected from small animal clinic ward of Madras Veterinary College during the period from November, 2024 to January, 2025 and all the faecal samples were subjected to DNA isolation followed by PCR was carried out by targeting VP2 capsid gene region of FPV along with vaccine strain as positive control. Among the 41 samples, 18 samples were found to be positive for FPV by PCR and revealed the case prevalence of 43.90 % from samples collected from in and around Chennai, Tamil Nadu. Further, to characterize the circulating the field FPV strains, the complete VP2 gene capsid protein was amplified and subjected to sequencing. The sequence analysis revealed that all the FPV strains were closely related with FPV strains from India and other country. Further, the antigenic relationship between vaccine and field FPV strains revealed that unique nucleotide and amino acid variation at genome level. This study provides insights on prevalence of feline panleukopenia virus in India and their antigenic relationship between vaccine and field FPV strains to prevent by vaccination.

KEYWORDS: *Feline Panleukopenia virus, FPV, VP2 gene, Molecular characterization, Prevalence.*

MOLECULAR SCREENING FOR NOVEL PORCINE CIRCOVIRUSES ASSOCIATED WITH REPRODUCTIVE FAILURE IN SWINE

J. Janaranjani, S. Parthiban*, K. Reshma, P. Raja, J. Vijay Anand, M. Parthiban

Department of Animal Biotechnology, Faculty of Basic Sciences,
Madras Veterinary College, Chennai-600 007
Tamil Nadu Veterinary and Animal Sciences University
**Corresponding author: S. Parthiban, Assistant Professor, Department of Animal Biotechnology, Madras* Veterinary College, Chennai-600007.
Author mail id: jaanujayakumar@gmail.com

ABSTRACT

Porcine Circovirus 3 (PCV3), associated with multisystemic inflammation, reproductive failure, and subclinical infections, and Porcine Circovirus 4 (PCV4), associated with respiratory and enteric infections, are novel porcine circoviral infections identified in many swine-producing countries. However, their molecular prevalence is not extensively explored in India. The present study examined 50 post-mortem tissue samples collected from aborted/stillborn piglets and swine that died with a history of systemic infections from Andhra Pradesh, Karnataka, Kerala, Pondicherry, Tamil Nadu, and Telangana states in southern India. All samples were processed for the molecular screening of novel Porcine Circovirus 3 (PCV3) and PCV4 using specific PCR assays. Molecular positivity for PCV3 was found to be 8% (n=4), whereas none of the samples tested positive for PCV4 in southern India. All four samples that tested positive for PCV3 were associated with reproductive failure and were negative for other porcine viruses linked to reproductive failure, such as PCV2, Porcine Parvovirus (PPV), and Classical Swine Fever Virus (CSFV). Representative PCV3 positive PCR amplicons were sequenced, revealing 651 nucleotides covering the complete capsid protein with deduced amino acids of 214. BLAST analysis confirmed the specific capsid sequence of the PCV3 genome. The PCV3 sequence identified in this study showed 99.85% homology with PCV3 isolates from China and Colombia. This study provides evidence of PCV3-associated reproductive failure in domestic pigs in southern India. Further studies focusing on capsid protein-based molecular characterization and examination of immunodominant regions of PCV3 are warranted.

KEYWORDS: *Molecular Screening, PCV3, PCV4, Reproductive failure, Swine.*

MICRONUTRIENT DELIVERY TO PLANTS USING NANOCOMPOSITE

Arputhaselvi R[1], Subbulakshmi C[1], Dr. Nalinkanth V Ghone[2*]

[1]Department of Chemical Engineering, Sri Sivasubramaniya Nadar College of Engineering, Kalavakkam, Chennai -603110, India

[*]Corresponding author mail id: nalinkanthv@ssn.edu.in

ABSTRACT

This study investigates the development and application of a nanocomposite fertilizer designed to enhance the productivity and quality of crops. The nanocomposite, which was thoroughly characterized using advanced analytical techniques, was applied via foliar spraying, a method that allows for direct nutrient delivery to the plant leaves, facilitating rapid absorption and improving nutrient use efficiency. Compared to traditional soil-based fertilization methods, this approach offers several advantages, including controlled nutrient release, reduced environmental loss, and enhanced bioavailability. These benefits led to improved plant growth, increased yield, and better overall crop quality. The findings highlight the potential of Nano fertilizers to transform agricultural practices by optimizing nutrient delivery, reducing fertilizer usage, and minimizing environmental impact, presenting a sustainable solution for boosting crop productivity.

KEY WORDS: *Nanocomposite, Foliar spraying, FT-IR*

SYNTHESIS OF DIMETHYL CARBONATE FROM SEQUESTERED CO_2 IN CLAY NANOTUBES

Giridaran S

Department of Chemical Engineering, Sri Sivasubramaniya Nadar College of Engineering, Kalavakkam, Chennai -603110, India

[*]Author mail id: giridaran2110674@ssn.edu.in

ABSTRACT

Carbon dioxide is a major greenhouse gas that traps heat in the earth's atmosphere, leading to global warming and changes in weather patterns. This results in more frequent and severe heat waves, droughts, floods, and storms, affecting ecosystems, agriculture, and communities. CO_2 capture by solid adsorbents offers benefits such as high adsorption capacity, easy recovery, high uptake efficiency under humid conditions, easy handling, and materials stability. In recent studies low-cost clay nanotube has been used for CO_2 adsorption and it was shown that Tetra ethylene pentamine modified clay nanotubes has enhanced the CO_2 adsorption by 60% when compared to CO_2 adsorption by clay nanotubes. Though clay nanotube has been shown as an economical versatile material for CO_2 adsorption. The regeneration of clay nanotube was always done by using thermal method (heating up to 120°C) which is very expensive and again leaves CO_2 footprint. Hence, there is a need for regeneration of clay nanotube at normal atmospheric temperature. We regenerated the clay nanotubes for further adsorption of CO_2 by utilising the adsorbed CO_2 on clay nanotube for synthesis of dimethyl carbonate (DMC) using methanol. DMC is widely used as methylation agent, a solvent and a fuel additive, hence successful outcome of this proposal has resulted in a clay nanotube-based carbon capture utilisation and storage technology (CCUS) which will result in production of valuable DMC.

KEY WORDS: *Nanotube, Dimethyl carbonate, FT-IR.*

REGENERATIVE DERMATOLOGICAL FORMULATION USING MEDICINAL PLANT EXTRACTS FORACNE SCAR AND STRETCH MARK HEALING

K. M. Amerdhavarsine

Final Year student, B.Tech. Biotechnology, Vivekanandha College of Engineering for Women,
Tiruchengode, Namakkal [DT]
Author mail id:amerdhavarsine2711@gmail.com

ABSTRACT

Acne scars and stretch marks are prevalent dermatological concerns, significantly impacting individuals' self-esteem and quality of life Conventional treatments, such as laser therapy and chemical peels, are often costly, short-term, and associated with adverse effects. This ongoing study focuses on developing a cost-effective, natural, and sustainable dermatological formulation utilizing medicinal plant extracts. The primary active ingredients include Aloe vera gel, *Centella asiatica* extract, and turmeric extract, recognized for their hydrating, collagen-stimulating, and anti-inflammatory properties. Supporting ingredients like beetroot powder, rose petal powder, and agar enhance the formulation's viscosity, stability, and sensory appeal, while sodium benzoate acts as a preservative. The extraction process employs solvent extraction methods, including ethanol-based techniques, ensuring efficient retrieval of bioactive compounds. The plant materials are proposed to be authenticated by the Botanical Survey of India to confirm their origin and quality for research purposes. The formulation will undergo rigorous testing for pH, viscosity, stability, microbial safety, and spreadability. Planned analyses include antioxidant tests, patch tests, and clinical efficacy trials, supported by sensory evaluations to assess user experience. Animal studies are being considered to validate the therapeutic potential before advancing to human trials. Preliminary insights suggest the formulation's efficacy in promoting skin regeneration, supported by its antioxidant activity and user acceptability. By integrating eco-friendly practices and scientifically validated methods, this project addresses the demand for sustainable, effective, and accessible skincare solutions. This work contributes to innovative dermatological treatments, aligning with global trends toward natural product development.

KEYWORDS: *Herbal Dermatology, Acne Scar Healing, Stretch Mark Reduction, Medicinal Plant Extracts, Sustainable Skincare.*

STUDIES ON THE ISOLATION OF BIOACTIVE COMPOUND FROM *CLEOME GYNANDRA* AND THEIR ANTIOXIDANT, *IN-VITRO* ANTI CANCER PROPERTIES AND *IN SILICO* MOLECULAR DOCKING

S. Keerthilakshmi[1], Shreya. S[1], Dr. R. Anila[2]

[1]Student at Department of Biotechnology at Rajalakshmi Engineering College.
[2]Associate Professor at Department of Biotechnology at Rajalakshmi Engineering College.
Author:pg.bt.2023@gmail.com

ABSTRACT

This study explores the isolation and characterization of bioactive compounds from *Cleome gynandra*, a plant valued for its traditional medicinal and nutritional uses. The research aims to extract, identify, and evaluate the therapeutic potential of these compounds, focusing on their antioxidant and anticancer properties. Extraction was performed using solvents like methanol, hexane, and water, followed by phytochemical screening to detect active constituents. The antioxidant activity of the extracts was assessed through DPPH and ABTS assays, providing insight into their ability to neutralize free radicals. The anticancer potential was investigated using in-vitro cytotoxicity studies against cancer cells, emphasizing apoptosis mechanisms, including mitochondrial disruption and oxidative stress. To understand molecular interactions, in-silico docking simulations analyzed how these compounds bind to cancer-related proteins. Advanced techniques such as thin-layer chromatography (TLC), column chromatography, and gas chromatography-mass spectrometry (GC-MS) were employed for compound purification and identification. Initial findings suggest that *Cleome gynandra* contains potent natural compounds with strong antioxidant activity and selective toxicity toward cancer cells, offering promise for its integration into cancer therapy. This research highlights the potential of *Cleome gynandra* asa natural source for developing safer and effective therapeutic agents.

KEYWORDS: *Cleomegynandra, Bioactive compounds, Antioxidant activity, Molecular docking, Phytochemical screening, Natural therapeutics.*

ANTI-MICROBIAL, ANTI-INFLAMMATORY ACTIVITY AGAINST URINARY TRACT INFECTIONS USING STROBILANTHES KUNTHIANA

Karishmadevi Y

Student, Master of Science, Department of biotechnology, Dr. MGR Educational and research institute, Chennai. Author:karishu1110@gmail.com

ABSTRACT

The shrub *Strobilanthes kunthiana*, also known as Neelakurinji, grows in the shola woods of South India's Western Ghats. Plant-based medicines have been utilised for treating human illnesses for eons, even before their history was written down. This is a bushy shrub that grows to a height of 1-2 meters in groups or clumps. The stems have distinct nodes and are upright and quadrangular in shape. The leaves have a coriaceous texture, a crenate-serrate edge, and an elliptic-ovate shape. The lower surface of the leaves is white-villous in between the veins. The present study has been designed to synthesise nanoparticles from the leaves of *Strobilanthes kunthiana*. The synthesised nanoparticles is characterised to check the biological activity using various assays. The assays like UV proved the synthesis of zinc nanoparticles. The FT-IR analysis showed the stretch bands between amide 1 and amide 2 groups. The TEM analysis revealed the crystalline structure and characteristics of synthesised zinc nanoparticles. It was found to be 50-200nm. The SEM analysis revealed the morphology of zinc oxide nanoparticles and the network formation appeared to be in spherical shape. The synthesised nanoparticles showed best and close results with antibiotics, which shows the synthesised nanoparticles extract shows high antibacterial activity. The nanoparticle synthesised from *Strobilanthes kunthiana* showed best antioxidant and anti-inflammatory activity. These results may justify the popular use of Zinc nanoparticles synthesized *Strobilanthes kunthiana* extract has antibacterial, anti-inflammatory property against the urinary tract infections. However in-order to evaluate the possible clinical application in therapy of urinary tract infections, further clinical trials are required.

KEYWORDS: *Strobilanthes kunthiana*, Zinc nanoparticles, Antioxidant, Anti-inflammatory, Urinary tract infection, Clinical trials.

ADVANCEMENTS IN MICROBIAL BIOTECHNOLOGY: INNOVATIONS, APPLICATIONS, AND SUSTAINABLE SOLUTIONS

B. Soumya, S. Rasika, M. Pravesihaa, S. Akshaya

Students, Department of Biotechnology, Arunai Engineering College, Thiruvannamalai.
Author:soumyabaskar2004@gmail.com

ABSTRACT

Microbial biotechnology has emerged as a transformative field with profound implications across various sectors. This presentation will delve into recent advancements in this field, exploring innovative technologies, diverse applications and their potential to address global challenges sustainably. We will discuss cutting edge techniques such as synthetic biology, CRISPR-Cas gene editing and metagenomics which are revolutionizing our understanding and manipulation of microbial systems. Furthermore, we will examine the multifaceted applications of microbial biotechnology including bioremediation, bioenergy production, industrial biotechnology and pharmaceutical development. The presentation will emphasize the crucial role of microbial biotechnology in developing sustainable solutions for environmental challenges such as pollution control and climate change mitigation. Finally, we will discuss the ethical considerations and future prospects of this rapidly evolving field, highlighting its potential to contribute to a more sustainable and prosperous future.

KEYWORDS: *Microbial biotechnology, Gene editing, CRISPR-Cas, Metagenomics, multifaceted applications.*

MOLECULAR PROFILING OF NON-SMALL CELL LUNG CANCER THROUGH SINGLE-CELL TRANSCRIPTOMIC DATA ANALYSIS AND *INSILICO* BIOLOGY.

SharathKumar R Munisankar, Anand Thiyagaraj*

Research scholar, Department of Genetic Engineering, School of Bioengineering, SRM Institute of Science and Technology, Kattankulathur, Chengalpet.
Author:rm.sharathkumar@gmail.com

ABSTRACT

Non-Small cell lung cancer (NSCLC) is the most prevalent subtype of lung cancer, contributing significantly to the global disease burden. This study employs single-cell RNA sequencing (scRNA-seq) to unravel the molecular complexity of NSCLC, deconvolute tumor heterogeneity, and identify cell types critical to its pathogenesis. The analysis begins with the clustering and annotation of cell types to identify populations driving tumor progression. Differential gene expression analysis will highlight cell type-specific marker genes, which will then undergo functional enrichment analysis to uncover their roles in key biological processes and pathways. Additionally, protein-protein interaction (PPI) networks will be constructed to delineate the molecular hubs and pathways that underlie NSCLC progression. Trajectory and pseudo time analyses will further elucidate the dynamic transcriptional changes and differentiation states of crucial cell populations. Gene regulatory network analysis will identify upstream transcription factors that modulate the expression of key biomarkers. To ensure robustness, findings will be validated across publicly available NSCLC RNA-seq datasets, integrating scRNA-seq-derived markers with bulk transcriptomic data. This comprehensive approach aims to reveal novel biomarkers and actionable targets, enhancing our understanding of NSCLC pathogenesis. The outcomes of this study are expected to inform precision oncology strategies, serving as a foundation for innovative diagnostic and therapeutic interventions in NSCLC.

KEYWORDS: *Non-small cell lung cancer, Single-cell RNA sequencing, Tumor heterogeneity, Biomarker discovery, Protein-protein interaction networks, Gene regulatory network.*

EVALUATING THE ANTI-INFLAMMATORY AND ANTIOXIDANT PROPERTIES OF PHYTOSTEROLS ISOLATED FROM *LAGENARIA SICERARIA* UNDER IN VITRO AND IN SILICO CONDITIONS

B. Harshitha, Dr. S. M. Fazeela Mahaboob Begum*

School of Life Sciences, B.S Abdur Rahman Crescent Institute of Science and Technology, Chennai 600048, Tamil Nadu, India

**Corresponding author: fazeelabegum@crescent.education*

ABSTRACT

Lagenaria siceraria (Cucurbitaceae) is a promising source of bioactive compounds since it has been traditionally used for the treatment of various diseases. People with inflammatory conditions like arthritis typically rely on medications like NSAIDs, corticosteroids, and DMARDs for treatment. However, these drugs However, these drugs tend to have more side effects. As a result, many are seeking alternative solutions in herbal remedies, which are perceived to be more effective, safer, and widely accepted. The present study aims to evaluate the immunomodulatory effect of *Lagenaria siceraria* extract both in vitro and in silico.

KEYWORDS: *Lagenaria siceraria, Bioactive compound, Phytosterols, Anti-inflammatory, Antioxidant properties.*

DEVELOPMENT OF NUTRACEUTICAL GUMMIES FOR SUPPORTING PCOS MANAGEMENT

Fathima M*, Ayesha Sithika S[1], Sanju Shree V[1], Sowmiya B[1]

Department of Biotechnology, Vivekanandha College of Engineering for Women
**Corresponding author: fathima@vcew.ac.in*

ABSTRACT

Polycystic Ovary Syndrome (PCOS) is a complex endocrine disorder affecting millions of women worldwide, characterized by hormonal imbalances, insulin resistance, and chronic inflammation. This study explores the development of nutraceutical gummies as a novel, user-friendly approach for PCOS management. These gummies incorporate evidence-based bioactive ingredients, including Agar – Agar, Lemon Juice and herbal extracts like" *Fagonia Indica*, Garcinia cambogia Desr, *Caesalpinia Bonducella*, *Trigonella Forum*, *Cinnamomum verum*, Mentha, Aleovera, Beetroot chosen for their synergistic roles in improving insulin sensitivity, regulating androgen levels, and alleviating systemic inflammation. The research focused on ingredient selection, formulation optimization, stability assessment, sensory evaluation, and preliminary efficacy testing. Results revealed that the developed gummies maintain the bioactivity of the ingredients, deliver a highly palatable consumer experience, and hold significant promise for improving metabolic, hormonal, and reproductive health in women with PCOS. This study highlights the potential of functional food products in addressing lifestyle and metabolic disorders. Future work will focus on clinical validation, scalability, and exploring consumer adoption to advance the application of nutraceutical gummies for personalized health management.

KEYWORDS: *PCOS, Nutraceuticals, gummies, Functional food, Hormonal imbalance, Inflammation, Women's health, Personalized nutrition.*

SELF-HEALING BIO-CONCRETE USING *BACILLUS SUBTILIS* ENCAPSULATED ZINC OXIDE NANOPARTICLES

Renganayagi C[1], Ranjani S[1], Revathy. J[2], S. Hemalatha[1] *

[1]*School of Life Sciences, B. S. Abdur Rahman Crescent Institute of Science and Technology,Vandalur, Chennai, 600048, Tamilnadu, India.*
[2]*Department of Civil Engineering, B. S. Abdur Rahman Crescent Institute of Science and Technology,Vandalur, Chennai, 600048, Tamilnadu, India.*
**Corresponding Author mail Id: hemalatha.sls@crescent.education*

ABSTRACT

Microbial-based self-healing bio-concrete is an innovative nanobiotechnological solution that enables buildings to autonomously repair cracks that develop due to environmental stressors including heavy rain and heat. Bio-concrete is composed of fine aggregates, coarse aggregates, ordinary Portland cement, and nanoparticles encapsulated bacteria, which enhance the structural integrity of concrete compared to conventional formulations. This study investigates the self-healing capabilities of bio-concrete by utilizing nanoparticles synthesized via the precipitation method, which are then used to encapsulate *Bacillus* sp. The nanoparticles were synthesized and characterized using various techniques, including UV-Vis spectroscopy, which confirmed the synthesis of ZnO nanoparticles with SPR peak at 370 nm. Additional characterizations were performed using Dynamic Light Scattering (DLS), Zeta Potential analysis, Fourier Transform Infrared (FTIR) spectroscopy, Field Emission Scanning Electron Microscopy (FESEM) with Energy Dispersive X-ray (EDAX) analysis, X-ray Diffraction (XRD), and Thermogravimetric Analysis (TGA). The nanoparticles-coated microbes exhibit enhanced efficiency and thermal stability, facilitating the endothermic reaction required for the precipitation of calcium carbonate when cracks form on the concrete surface. This research aims to evaluate the potential of novel *Bacillus* strains encapsulated in zinc oxide nanoparticles to precipitate calcium carbonate and self-healing processes. The findings from this study will contribute to a deeper understanding of self-healing concrete, offering promising prospects for sustainable construction practices and green cement production.

KEYWORDS: *Microbial-based self-healing, Bio-concrete, Zinc oxide nanoparticles, Calcium carbonate precipitation, Bacillus strains, Sustainable construction.*

TRANSPARENCY OF NUTRITIONAL LABELLING OF NANO FOODS

Rashwitha C S Poojary[1], Jyosthna Devi Kuchipudi*

[1]M.Sc. Student, Department of Food, Nutrition and Dietetics, School of Allied Healthcare and Sciences. Jain Deemed to be University, Bengaluru.
**Assistant Professor, Department of Food, Nutrition and Dietetics, School of Allied Healthcare and Sciences. Jain Deemed to be University, Bengaluru.*
**Corresponding Author mail Id: jyosthna.k@jainuniversity.ac.in*

ABSTRACT

The advent of nanotechnology in food production has transformed the sector by improving food quality, safety, and functionality. Nanofoods, which integrate nanoscale materials or structures, enhance texture, flavour, nutrient delivery, and shelf life. Their expanding presence in the worldwide market requires open and precise nutritional labelling to guarantee customer trust, safety, and informed decision-making. This analysis examines the existing issues and prospects regarding the transparency of nutritional labelling for nanofoods. The integration of nanoscale additives, including nanoencapsulated vitamins, minerals, and bioactive substances, markedly enhances nutrient bioavailability. The nanoscale characteristics of these components prompt concerns over their safety, bioaccumulation, and long-term health effects. Regulatory frameworks governing nanofoods and labelling regulations are different between areas, leading to consumer confusion and scepticism. Clear labelling, which includes the explicit identification of nanoscale substances, their functional advantages, and potential hazards, is essential for enhancing consumer trust and adherence to food safety regulations. Additionally, the ethical ramifications of nanofood production and labelling are examined, highlighting the necessity for standardised global regulations. This analysis emphasises the necessity of interdisciplinary cooperation among regulatory agencies, food technologists, and consumer advocacy organisations to develop comprehensive labelling standards for nanofoods. Future initiatives should concentrate on amalgamating developments in nanoscience with regulatory frameworks to mitigate safety issues while optimising the nutritional benefits of nanofoods.

KEYWORDS: *Nanotechnology, Nutritional labelling, Transparency, Consumer awareness, Nano foods.*

POLYHERBAL NANOEMULSION: A NOVEL STRATEGY FOR SUSTAINABLE MANAGEMENT OF FUNGAL INFECTIONS IN *SACCHARUM OFFICINARUM*

Divyashree V, Ranjani S, S. Hemalatha*

School of Life Sciences, B. S. Abdur Rahman Crescent Institute of Science and Technology, Vandalur, Chennai, 600048, Tamilnadu, India.

**Corresponding Author mail Id: hemalatha.sls@crescent.education*

ABSTRACT

Sugarcane (*Saccharum officinarum*) is a vital cash crop globally; however fungal pathogens significantly impact its yield. Sugarcane is a most important crop for sugar production and fungal diseases in sugarcane contribute to 15 to 30 % of loss. Current management strategies often lack sustainability and efficacy. This study mainly aims to develop a Polyherbal Nanoemulsion to effectively manage fungal infections in *Saccharum officinarum.* This study contributes to investigating the green synthesized silver nanoparticles from different plant extracts and medicinal oil. Antifungal study of nanoemulsion will be carried out in fungal pathogens isolated from infected sugarcane. This study helps to develop polyherbal nanoformulation to control and manage fungal pathogens in sugarcane and help farmers to improve their economy.

KEYWORDS: *Saccharum officinarum, Silver Nanoparticles, Polyherbal Nanoemulsion, Anti-Fungal Activity, Sustainable agriculture.*

GUT MICROBIOTA IN ACTION: PROBIOTICS AS NUTRITIONAL ADVANCEMENT

Samyama S Gowda[1], Jyothsna Devi Kuchipudi*

[1]MSc. Student, Area of Food Nutrition and Dietetics, School of Allied Healthcare & Sciences, Jain Deemed to be University, Bengaluru.
**Assistant Professor, Area of Food Nutrition, and Dietetics, School of Allied Healthcare & Sciences, Jain Deemed to be University, Bengaluru.*
**Corresponding Author mail Id: jyothsna.k@jainuniversity.ac.in*

ABSTRACT

The gut microbiota, a complex ecology of billions of bacteria inhabiting the gastrointestinal tract, is essential for host metabolism, immunity, and overall health. This review examines the molecular mechanisms by which probiotics affect nutrition, emphasizing species such as *Lactobacillus rhamnosus, Bifidobacterium longum, Lactobacillus acidophilus, Saccharomyces boulardii, Streptococcus thermophilus, and Bacillus coagulans.* These probiotics facilitate the degradation of complex carbohydrates into short-chain fatty acids (SCFAs), such as acetate, propionate, and butyrate, which serve as energy substrates and maintain gut epithelial integrity. Furthermore, *Bifidobacterium breve* and *Lactobacillus casei* enhance the bioavailability of micronutrients, including calcium, iron, and zinc, by diminishing anti-nutritional factors and enhancing intestine absorption. *Streptococcus thermophilus* and *Lactobacillus delbrueckii p*roduce vital B-group vitamins, such as folate, riboflavin, and vitamin B12, which are required for cellular metabolism. Probiotics demonstrate anti-inflammatory properties, decreasing intestinal permeability and alleviating conditions such as irritable bowel syndrome, lactose intolerance, and inflammatory bowel disease. This review highlights the diverse contributions of probiotics to nutritional research and their potential to tackle global health issues, such as malnutrition, metabolic diseases, and micronutrient shortages. Future research must concentrate on strain-specific effects, probiotic-genome interactions, and the creation of next-generation probiotics designed for precision nutrition and medicinal uses.

KEY WORDS: *Gut microbiota, Probiotics, Lactobacillus rhamnosus, Micronutrient bioavailability.*

BIOMARKER BASED PERSONALIZED NUTRITION

Vishnu Abirami.s[1], Jyothsna Devi Kuchipudi *

[1]M.Sc. Student, Area of Food, Nutrition and Dietetics, School of Allied Healthcare & Sciences. Jain Deemed to be University, Bengaluru.

** Assistant Professor, Area of Food, Nutrition and Dietetics, School of Allied Healthcare & Sciences. Jain Deemed to be University, Bengaluru*

**Corresponding Author mail Id: jyothsna.k@jainuniversity.ac.in*

ABSTRACT

Biomarkers serve as pivotal tools in personalized nutrition, enabling dietary recommendations styled to individual biological profiles. They are objective and quantifiable indicators of biological processes that are classified into exposure and disease markers. While exposure markers reflect dietary interactions and disease markers assist in diagnosing and managing nutrition related health conditions. Omics technologies- "Genomics, Proteomics, Metabolomics and Microbiome analysis"- drive advancements in personalized nutrition. Genomics highlights genetic predispositions to nutrient responses, providing insights into individual risks for conditions such as obesity or diabetes. Proteomics reveals protein expression patterns, linked to cellular responses to specific nutrients or toxicities. Metabolomics assesses metabolic responses, enabling biomarkers for early disease detection and microbiome analysis elucidates the guts' role in nutrient metabolism. These biomarkers complement traditional dietary assessment methods by providing objective measures of nutrient intake, identifying deficiencies, and enabling precision dietary guidance. Despite their promise, challenges such as high cost, data complexity, and ethical concerns about genetic testing hinder widespread adoption. Biomarker driven nutrition offers significant benefits, including prevention of nutrient related diseases, improved adherence to dietary plans and an evidence based public health strategy. By overcoming Implementation barriers, biomarker based personalized nutrition holds the potential to revolutionize dietary recommendations, optimize health outcomes, and shape public health policies, particularly in diverse populations.

KEYWORDS: *Biomarkers, Omics technologies, Personalized Nutrition, Dietary assessment Indian public health, Nutrigenomics.*

PREVENTION OF BIOFILM BY ANTIMICROBIAL COATINGS IN CATHETER

S. Madhumitha, S.Savitha, V. Thivija

Department of Biotechnology, Vivekanandha College of Engineering for Women
Author: savithasekar630@gmail.com

ABSTRACT

Biofilm formation on urinary catheters is a major cause of catheter-associated urinary tract infections (CAUTIs), posing significant health risks and treatment challenges. Herbal coatings have gained attention as a natural and sustainable approach to prevent biofilm development. Derived from plant-based compounds with antimicrobial, anti-inflammatory, and anti-adhesive properties, herbal coatings inhibit microbial adhesion and disrupt biofilm formation without promoting resistance. Commonly used herbal plants which is easily available. These coatings offer advantages such as biocompatibility, reduced toxicity, and minimal environmental impact. Herbal-coated urinary catheters present a novel and eco-friendly solution for reducing CAUTIs, improving patient health, and addressing the global concern of antimicrobial resistance. The main objective of this project is to evaluate the effectiveness of antimicrobial coatings against common urinary pathogens. The ultimate aim is to design anti biofilm coatings for stability through in vitro studies.

KEYWORDS: *Catheter, Urinary tract infections, Antimicrobial properties, Biocompatibility*

EXPLORING THE BIOMEDICAL APPLICATIONS OF PINK OYSTER MUSHROOM MEDIATED NANOMATERIAL

Sharfia Naaz Jabbar,Ranjani S and S. Hemalatha*

School of Life Sciences, B. S. Abdur Rahman Crescent Institute of science and technology, Vandalur, Chennai, 600048, Tamil Nadu, India.

**Corresponding author: hemalatha.sls@crescent.education*

ABSTRACT

In this study, we synthesized nanomaterial by using *Pleurotus djamor* (Pink Oyster Mushrooms) extract, which is rich in bioactive compounds with antioxidant, antimicrobial and anti-apoptotic properties. This nanomaterial was characterized using techniques such as UV-Vis spectroscopy, FTIR, DLS, zeta potential, XRD and FESEM-EDAX to confirm their formation and stability. The antifungal potential of the synthesized nanomaterial was evaluated in *Candida* species and exhibited significant antifungal efficacy, by inhibiting growth and biofilm formation. These findings highlight the potential of *Pleurotus djamor*-mediated nanomaterial as an effective alternative for combating antifungal resistance and managing infections caused by *Candida* species. This study paves the way for candida infection and its associated co infection management.

KEYWORDS: *Pink Oyster Mushrooms, Silver Nanoparticles, Antifungal activity.*

MARINE SEAWEED MEDIATED SELENIUM NANOPARTICLES AND ITS EFFECTS ON ALZHEIMER'S DISEASE

Jameema S, Ranjani S, Architha V, S.Hemalatha *

School of Life Sciences, B. S. Abdur Rahman Crescent Institute of Science and Technology, Vandalur, Chennai, 600048, Tamilnadu, India.
**Corresponding author: hemalatha.sls@crescent.education*

ABSTRACT

Alzheimer's disease (AD)is caused by complex changes in the brain that start years before symptoms appear. AD is the most common cause of dementia in older adults. Dementia is the loss of the ability to reason, learn new skills, and plan and prioritize. This study focuses on the green synthesis of selenium nanoparticles (SeNPs) using an aqueous extract from a bioactive-rich marine seaweed abundantly found along the southern coasts of India. Seaweed is known for its antioxidant, anti-inflammatory, and neuroprotective properties; hence seaweed offers a sustainable, eco-friendly alternative to conventional chemical methods. Characterization was performed using techniques including FESEM, XRD, EDAX, DLS, Zeta Potential, and FTIR, confirming their uniform size and stability. Selenium, a vital micronutrient, combats oxidative stress, reduces inflammation and enhances cellular defense. It addresses key factors in Alzheimer's disease, making it ideal for biomedical applications. The therapeutic efficacy of SeNPs is being evaluated in an Alzheimer's-induced zebrafish model, focusing on cognitive improvements, oxidative stress reduction, and neuroprotection. This scalable green synthesis approach highlights SeNPs as a sustainable, effective therapeutic solution for Alzheimer's disease, with significant potential for future clinical applications.

KEYWORDS: *Selenium nanoparticles (SeNPs), Green synthesis, Alzheimer's disease, Zebrafish model.*

GREEN REMEDY THROUGH THE FUNCTIONALIZED ORGANOCLAY TO REMOVE BISPHENOL A

Hephziba X, Architha V, S. Hemalatha*

School of Life Sciences, B. S. Abdur Rahman Crescent Institute of Science and Technology, Vandalur, Chennai, 600048, Tamilnadu, India.

**Corresponding author: hemalatha.sls@crescent.education*

ABSTRACT

The increasing presence of organic pollutants in water sources poses a severe threat to both the environment and human health. Bisphenol A (BPA), a synthetic compound widely used in the manufacturing of plastics and resins, is one of the most concerning contaminants due to its endocrine-disrupting effects and potential to cause reproductive and developmental disorders. Traditional methods for removing BPA from water, such as filtration and chemical treatments, are often inefficient and costly. Organoclays, which are modified clays with organic compounds, have gained significant attention for their potential in adsorbing organic contaminants due to their enhanced surface area and hydrophobicity. In this study, hybrid organoclay by functionalizing bentonite with zinc oxide (ZnO) and polyethylene glycol (PEG) was prepared. For characterization UV-Spectrometry and Zeta potential analysis was done which implies the functionalization has more stability and adsorption capacity. Further characterization will be performed using techniques such as Fourier Transform Infrared (FTIR) spectroscopy, Field Emission Scanning Electron Microscopy (FESEM) with Energy Dispersive X-ray (EDAX) analysis, X-ray Diffraction (XRD), Brunauer-Emmett-Teller (BET) and Thermogravimetric Analysis (TGA).Toxicity analysis was performed in *V.radiata* to assess the material's properties and safety. Initial experiments showed improved adsorption capacity of this functionalized bentonite, but further optimization is required for effective dye degradation, for this purpose modification with CTAB will be experimented. This will be evaluated for its effectiveness in removing BPA from contaminated water. This research contributes to the development of efficient and sustainable materials for wastewater treatment, with potential applications in industrial effluent management and Bioremediation.

KEY WORDS: *Organoclay, Bisphenol A, Functionalized Bentonite, Vignaradiata, Bioremediation.*

SYNTHESIS AND CHARACTERIZATION OF SELENIUM NANOPARTICLES USING *ULVA FASCIATA* FOR ANTIDIABETIC APPLICATIONS

Afreen S, Ranjani S, Archita V and S. Hemalatha*

School of Life Sciences, B.S. Abdur Rahman Crescent Institue of Science and Technology, Vandalur, Chennai 600048, Tamil Nadu, India

**Corresponding author: hemalatha.sls@crescent.education*

ABSTRACT

Green synthesis of selenium nanoparticles (SeNPs) offers a sustainable and innovative approach for biomedical applications. In this study, *Ulva fasciata Delile*, a marine macroalga, was used as a biogenic source to synthesize SeNPs, exploiting its phytochemical properties for reduction and stabilization. The nanoparticles were characterized using FESEM, XRD, DLS, zeta potential, UV-Vis, GC-MS, and FTIR to elucidate their structural and physicochemical features. Their antidiabetic activity was assessed through α-amylase and α-glucosidase inhibition assays, while antioxidant potential was evaluated using DPPH assay. Toxicity and biocompatibility were tested on zebrafish (*Danio rerio*). This study highlights the therapeutic potential of *U. fasciata*-derived SeNPs, offering a green and effective solution for diabetes management and oxidative stress mitigation.

KEYWORDS: *Green synthesis, Selenium Nanoparticles, Antidiabetic, Antioxidant, Danio rerio.*

INVESTIGATING THE NEUROPROTECTIVE POTENTIAL OF MEDICINAL PLANTEXTRACT: MODULATION OF HSP70 MEDIATED EXPRESSION OF GENES INVOLVED IN CAUSING ALZHEIMER'S DISEASE

Aparna.L.N, Architha V, S. Hemalatha*

School of Life Sciences, B. S. Abdur Rahman Crescent Institute of science and technology, Vandalur, Chennai, 600048, Tamil Nadu, India.
Corresponding author: hemalatha.sls@crescent.education

ABSTRACT

Cognitive impairment and amyloid beta (Aβ) toxicity are hallmarks of Alzheimer's disease (AD), a progressive neurodegenerative disorder with limited treatment options and significant adverse effects associated with existing therapies. The search for alternative treatments has directed attention toward plant-based remedies due to their abundance of bioactive compounds with antioxidant, anti-inflammatory, and neuroprotective properties. In this study, we evaluate the therapeutic potential of a medicinal plant extract in modulating heat shock proteins (HSP70), which play a critical role in protein homeostasis and protection against neurodegeneration. Zebrafish embryos, an established model for neurodegenerative disease studies due to their genetic similarity to humans and rapid developmental cycles, were employed to investigate the effects of the medicinal plant extract. The study focuses on assessing both the developmental toxicity of the extract and its efficacy in mitigating Aβ-induced cellular stress through HSP70 modulation. The findings aim to elucidate the molecular mechanisms underpinning the extract's neuroprotective action, offering insights into its safety profile and potential application as a novel therapeutic agent for managing AD. This research contributes to the growing body of evidence supporting the integration of natural compounds in neurodegenerative disease therapy.

KEYWORDS: *Cognitive impairment; Amyloid beta (Aβ) toxicity; Alzheimer's disease; Heat shock proteins; Zebrafish embryos.*

GREEN NANOEMULGEL FOR EFFECTIVE ACNE TREATMENT

Maahin AafiyaR, Ranjani Sand S. Hemalatha *

School of Life Sciences, B. S. Abdur Rahman Crescent Institute of Science and Technology, Vandalur, Chennai, 600048, Tamilnadu, India.
**Corresponding author: S. Hemalatha,*
Corresponding author mail ID: hemalatha.sls@crescent.education

ABSTRACT

Nanotechnology has emerged as a promising field in the development of skincare formulations and the demand for natural and safe production of skincare and biomedical products has driven interest in plant-based synthesis of nanoparticles. This study aims to explore the green synthesis of silver nanoparticles (AgNPs) by using an aqueous extract of *Salvia rosemarinus.* The synthesized AgNPs was characterized through various techniques including UV-Vis spectroscopy, Dynamic light scattering (DLS), Zeta potential analysis, FTIR spectroscopy, FESEM, EDAX and XRD. These nanoparticles were incorporated into a nanoemulgel formulation containing lavender (*Lavandula angustifolia*) essential oil to enhance its wound healing and antiseptic properties. Antimicrobial assays will be conducted *in vitro* in acne-causing bacteria using various assays. This study aims to develop a promising plant-derived green nanoemulgel that combines the benefits of nanoparticles and medicinal oil for slow and sustainable release.

KEYWORDS: *Nanotechnology, Silver Nanoparticles, Nanoemulgel, Nanoemulsion, Acne-causing bacteria.*

GREEN NANOFUNGICIDE TO CONTROL FOOD CONTAMINATING PLANT PATHOGENS

Ranjani S, S. Hemalatha*

School of Life Sciences, B. S. Abdur Rahman Crescent Institute of Science and Technology, Vandalur, Chennai, 600048, Tamilnadu, India.
**Corresponding author: S. Hemalatha,*
Corresponding author mail ID: hemalatha.sls@crescent.education, ranjani.sls@crescent.education

ABSTRACT

Phyto-pathogens affect agricultural crops and reduce the yield worldwide. Agrochemicals including fungicides have been widely used to control the plant fungal pathogens. Moreover, these agrochemicals are lethal to the beneficial organisms. Hence, there is a pressing need to develop potent and cost effective antifungal agents. The synthesis of green nanofungicide using green approach and its application in biological fields is a burgeoning field of nanotechnology. *Aspergillus niger* produces the potent mycotoxins called ochratoxins that contaminates food grains. *Fusarium graminearum* causes Fusarium head blight or scab of wheat and barley. *Fusarium graminearum* produces group of vomitoxins which contaminates wheat and barley and cause liver diseases. *Fusarium udum* causes wilt disease in pigeon pea plants. All these pathogens spread through the irrigation and damage growth, development and yield in food crops. The above selected fungal plant pathogens are commercially important and cause various diseases on vegetables, fruits, cereals, and cereal crops. In this study, we have synthesized green nanofungicide as an alternate to the commercial fungicides to control plant fungal pathogens such as *Aspergillus niger, Fusarium graminearum* and *Fusarium udum.* The green nanofungicide showed broad-spectrum antifungal activity against the phyto-pathogens by inhibiting the radial growth. Green nanofungicide could be used as effective antifungal agents against harmful phytopathogenic fungi. Our results suggests that green nanofungicide can be utilized as an alternative for fungicides to prevent diseases in food crops and to eliminate the contamination of food grains to eradicate food poisoning and its associated health effects.

KEYWORDS: *Green Nanofungicide, Antifungal activity, Phytopathogens.*

BIOMATERIAL SYSTEM OF*TYLOPHORA INDICA*(AQUEOUS EXTRACT) MEDIATED BIMETALLICSILVER – SELENIUM NANOPARTICLES LOADED HYDROGEL FOR LOCALIZED DIABETIC WOUND HEALING AND CANCER THERAPY

Rasitha Arafa Rafiudeen, Ranjani S, and Hemalatha S*

School of Life Sciences, B. S. Abdur Rahman Crescent Institute of science and technology, Vandalur, Chennai, 600048, Tamil Nadu, India.
**Corresponding author: S. Hemalatha,*
Corresponding author mail ID: hemalatha.sls@crescent.education,

ABSTRACT

Nanobiotechnology is a transformative field in therapeutics with innovative approaches to drug delivery, wound healing, and cancer therapy. This research focuses on the green synthesis of silver (Ag), selenium (Se), and bimetallic silver-selenium (Ag-Se) nanoparticles using the aqueous extract of *Tylophora indica*, an Indian medicinal plant known for its therapeutic properties. The synthesis of these nanoparticles was confirmed through UV-Visible spectroscopy producing characteristic peaks at 422 nm for Ag, 258 nm for Se, and 350 nm for Ag-Se nanoparticles. Further characterization will be carried out using techniques such as DLS, zeta potential analysis, FESEM, EDAX, XRD, and FTIR. Phytochemical screening of the extract is given using GC-MS, and identified phytocompounds will be evaluated through molecular docking for their potential interactions with wound healing and cancer target proteins. The synthesized nanoparticles will be studied for its antimicrobial, antioxidant, anti-inflammatory, and anticancer properties to support their application in diabetic wound healing and cancer therapies. Later, hydrogel will be formulated from the plant extract by incorporating the Ag-Se nanoparticles. This forms a hybrid biomaterial system which will be assessed for its physicochemical properties, biocompatibility, and therapeutic efficacy in wound healing assays and cancer studies. This study will define the potential of plant-mediated bimetallic nanoparticles and their hydrogel formulations in developing sustainable and localized therapeutic solutions for chronic conditions.

KEYWORDS: *Tylophora indica, Biomaterial systems, Green synthesis, Silver-Selenium Nanoparticles, Hydrogel, Diabetic Wound Healing, Cancer Therapy, Nanobiotechnology.*

CONTROLLING DENTAL PATHOGENS BY NANOMATERIALSYNTHESIZED FROM AERIAL ROOTS OF *FICUS BENGHALENSIS*

Aasiyah Tanveer Ahamed, Ranjani S and S. Hemalatha*

School of Life Sciences, B. S. Abdur Rahman Crescent Institute of Science and Technology, Vandalur, Chennai, 600048, Tamil Nadu, India.
**Corresponding author: S. Hemalatha,*
Corresponding author mail ID: hemalatha.sls@crescent.education,

ABSTRACT

A common disease faced by both children and adults worldwide is dental caries.Several virulence genes are associated with the formation of dental caries, which occur through various metabolic pathways. The virulence genes are responsible for biofilm formation, adhesion, polysaccharide production, and carbohydrate cleavage with acid production. This study focuses on the synthesis of silver nanoparticles by using the aqueous extract of aerial roots of *Ficus benghalensis* and its effects on the dental pathogens namely *Streptococcus mutans* and *Enterococcus faecalis. Ficus benghalensis* possesses antimicrobial properties making it a suitable candidate against dental pathogens. The synthesized silver nanoparticles were characterized by using UV-Vis spectroscopy, Fourier Transform Infrared (FTIR) spectroscopy, Dynamic Light Scattering (DLS), Zeta Potential analysis, X-ray Diffraction (XRD), and Field Emission Scanning Electron Microscopy (FESEM) with Energy Dispersive X-ray (EDAX) analysis. Dental pathogenic strains, *S. mutans* and *E. faecalis* were tested for antibacterial activity to determine the minimum inhibitory concentration (MIC) and minimum bactericidal concentration (MBC) of nanoparticles. The biofilm assay confirms the anti-biofilm potential of the silver nanoparticles. The study suggests that the silver nanoparticles using the aqueous extract of aerial roots of *Ficus benghalensis* possess tremendous potential in developing dental products like dental caps, fillers, and toothpaste.

KEYWORDS: *Green synthesis, Silver nanoparticles, Ficus benghalensis, Dental caries, Dental pathogens, S. mutans, E. faecalis, Biofilm.*

BIOSENSOR-ENABLED SMART PACKAGING: TRANSFORMING FOOD SAFETY AND WASTE REDUCTION

Jyothsna Devi Kuchipudi[1*],Shweta Singh[1]

[1]Ms. Jyothsna Devi Kuchipudi, Assistant Professor, Area of Nutrition and Dietetics, School of Allied Healthcare and Sciences, JAIN (Deemed to be University), Bangalore-560066, Karnataka, India.*

[1]Ms. Shweta Singh, Assistant Professor, Area of Nutrition and Dietetics, School of Allied Healthcare and Sciences, JAIN (Deemed to be University), Bangalore-560066, Karnataka, India.

**Corresponding author: Jyothsna Devi Kuchipudi*

Corresponding author mail ID: jyothsnadevikuchipudi@gmail.com

ABSTRACT

The integration of sensor technologies in food packaging has become a revolutionary method for ensuring food safety, minimizing waste, and enhancing sustainability. This review offers an extensive examination of several biosensors, encompassing electrochemical, optical, and piezoelectric sensors, and their functions in assessing essential food quality parameters. Electrochemical sensors accurately identify biochemical markers, including pH and enzyme activity, facilitating the early detection of deterioration. Optical sensors utilizing fluorescence or colorimetric alterations provide non-invasive, real-time surveillance of volatile organic compounds (VOCs) and gas emissions. Piezoelectric sensors detect microbial contamination with great sensitivity by exploiting variations in mechanical vibrations. These sensors were incorporated into biodegradable packaging materials to create environmentally sustainable, multipurpose smart packaging. Research findings illustrate the efficacy of these sensors in monitoring rotting and contamination in perishable food samples across various storage settings. This review emphasizes the integration of modern sensor technology with sustainable packaging, facilitating the development of next-generation food safety solutions.

KEYWORDS: *Biosensors, Electrochemical sensors, Optical sensors, Piezoelectric sensors, Smart packaging, Food safety.*

DRYNARIA QUERCIFOLIA MEDIATED BIOCOMPATIBLE ZINC NANOPARTICLES FOR COSMETIC APPLICATIONS

Chitra S, Ranjani S, S.Hemalatha

School of Life Sciences, B. S. Abdur Rahman Crescent Institute of Science and Technology, Vandalur, Chennai, 600048, Tamil Nadu, India.

**Corresponding author: S. Hemalatha,*

Corresponding author mail ID: hemalatha.sls@crescent.education,

ABSTRACT

The increasing demand for natural and sustainable products in cosmetics has led to the exploration of plant-based synthesis of nanoparticles. This study investigates the green biosynthesis of zinc nanoparticles (ZnNPs) using the aqueous extract of *Drynaria quercifolia*, a tuber known for its medicinal properties. The extraction of tuber was followed by green synthesis of nanoparticles, characterized by using UV-Vis spectroscopy, where a peak at 322 nm confirmed the formation of ZnNPs. Further characterization was performed using techniques including Dynamic Light Scattering (DLS), Zeta Potential analysis, Fourier Transform Infrared (FTIR) spectroscopy, Field Emission Scanning Electron Microscopy (FESEM) with Energy Dispersive X-ray (EDAX) analysis, and X-ray Diffraction (XRD). Candida species (yeast) cause infections on skin, lips eyes etc hence cosmetic industries are focussing on developing products that can eradicate yeast infection. Antifungal activity of the nanoparticles was assessed in four *Candida* species and showed potential antifungal properties with inhibition in biofilm formation. The findings suggest that *Drynaria* derived ZnNPs (DqZnNps) possess significant promise in the development of novel cosmetics with antifungal activity.

KEYWORDS: *Drynaria quercifolia*, Zinc Nanoparticles, Green synthesis, Antifungal Activity, *Candida* sp., Cosmetic Applications.

FUNCTIONAL DIFFERENCES IN HIV-1 AND HIV-2 TAT PROTEIN

Anitha Thulasisingh[1*],Luke Elizabeth Hanna[2], Vibha Murali [3]

[1*]*Associate Professor, Department of Biotechnology, Rajalakshmi Engineering College, Thandalam, Chennai – 602 105, India.*

[2]*Scientist "F" & HOD, Department of Virology & Biotechnology, ICMR- National Institute for Research in Tuberculosis, Chennai.*

[3]*M.Tech Biotechnology, Department of Biotechnology, Rajalakshmi Engineering College, Thandalam, Chennai – 602 105, India.*

ABSTRACT

HIV-1 and HIV-2 are two distinct types of human immunodeficiency viruses that differ markedly in their pathogenesis and disease progression. HIV-2 is associated with slower disease progression, reduced transmission rates, and lower plasma viral RNA levels compared to HIV-1. A key determinant of these differences lies in the Tat protein, a viral transactivator essential for viral replication and infectivity. HIV-2 exhibits significantly lower levels of Tat mRNA and protein expression, resulting in a reduced viral set point and slower replication. In contrast, HIV-1 shows elevated Tat expression, correlating with higher viremia and faster disease progression. Additionally, the rare codon usage observed in the Tat gene of HIV-2 may further impact its expression and functional efficiency, setting it apart from HIV-1. This study investigates the mechanistic basis of the divergent disease progression between HIV-1 and HIV-2, emphasizing the role of Tat. Furthermore, it examines the effects of point mutations within conserved and non-conserved regions of the functional domains of Tat. By elucidating the molecular and functional differences in Tat expression and activity, this research aims to shed light on the factors contributing to slower disease progression and reduced transmission rate in HIV-2. The findings may serve as a foundation for developing innovative therapeutic strategies targeting Tat to modulate disease progression and improve outcomes in HIV-infected individuals.

KEYWORDS: *HIV, Tat protein, disease progression, viral set point.*

COMPLEMENT PROTEINS AS PROGNOSTIC BIOMARKERS IN TUBERCULOSIS

Dr. G. Carlin Geor Malar[1*], Dr. Nancy Hilda Joseph[2], Azeeba Nousin A[1]

[1]*Department of Biotechnology, Rajalakshmi Engineering College, Thandalam, Chennai – 602 105, India.*

[2]*Department of Virology& Biotechnology, ICMR-NIRT, Chennai.*

ABSTRACT

Tuberculosis (TB) remains a leading global health challenge, causing significant morbidity and mortality despite advancements in diagnostics and therapeutics. The complement system, a pivotal component of the innate immune response, has emerged as a critical player in host defence against *Mycobacterium tuberculosis* (MTB). Dysregulation of complement pathways has been linked to disease progression, underscoring the potential of complement proteins as biomarkers for TB diagnosis and prognosis. This study investigates the diagnostic and prognostic utility of complement proteins in TB through a longitudinal analysis of plasma samples from 80 TB patients at three critical time points—baseline, mid-treatment (Month 2 or 6), and end of treatment—compared with 20 healthy controls. Two comprehensive panels of complement proteins targeting the classical, lectin, and alternative pathways were employed to identify key proteins associated with active TB, treatment response, and clinical outcomes. Preliminary findings highlight the ability of complement proteins to distinguish TB patients from healthy individuals, track treatment efficacy, and potentially predict outcomes. These results provide valuable insights into the immunological landscape of TB, paving the way for the development of novel diagnostic and prognostic strategies. This work emphasizes the utility of complement proteins as reliable biomarkers, offering promise for more effective TB management and control.

KEYWORDS: *Tuberculosis (TB), Complement system, complement proteins, Diagnostic biomarkers, Prognostic biomarkers.*

SELENIUM NANOPARTICLES-LOADED NIOSOMES: A DUAL-FUNCTIONAL APPROACH FOR ENHANCED PROBIOTIC DELIVERY AND ANTIMICROBIAL THERAPY

Niranjana. U, Ranjani. S, S. Hemalatha*

School of Life Sciences, B. S. Abdur Rahman Crescent Institute of Science and Technology, Vandalur, Chennai, 600048, Tamil Nadu, India.

**Corresponding author: S. Hemalatha,*

Corresponding author mail ID: hemalatha.sls@crescent.education,

ABSTRACT:

Gut health significantly impacts overall well-being, influencing digestion, immunity, and mental health. Maintaining a balanced gut microbiota is critical for preventing gastrointestinal disorders, metabolic conditions, and cancers. A novel approach involves synthesizing selenium nanoparticles-loaded niosomes by using polyherbal extracts. These non-ionic surfactant-based vesicles enhance the stability and delivery of probiotics, ensuring their survival through harsh gastrointestinal conditions including acidic pH and bile salts. Selenium nanoparticles (SeNPs), known for their antioxidant properties, high bioavailability, and low toxicity, further enhance immune function and exhibit potent antimicrobial and anticancer activities due to the enhanced permeability and retention effect of niosomes. This study explores the polyherbal synthesis of SeNPs, characterized by using UV-Vis spectrophotometry, FTIR analysis, Zetasizer DLS, FESEM-EDAX, and XRD. The nanoparticles demonstrated significant anti-diabetic, anti-inflammatory, antioxidant, and antibacterial properties, particularly controlling food-borne pathogens. This dual functionality of both probiotic delivery for gut health and antimicrobial/anticancer efficacy offers a versatile therapeutic platform. Future studies focus on investigating the biologically active compounds in polyherbal extracts to evaluate their pharmacological potential.

KEYWORDS:

Gut Health, Selenium Nanoparticles, Niosomes, Polyherbal Extracts, Probiotics, Antioxidant, Antimicrobial, Anticancer, Drug Delivery, Gastrointestinal Stability.

MICROFLUIDICS BASED GREEN SYNTHESIS OF SILVER NANOPARTICLES TO CONTROL BACTERIAL PATHOGENS CAUSING DENTAL CARIES

M. Annish Shabiya, Ranjani S, S. Hemalatha *

School of Life Sciences, B. S. Abdur Rahman Crescent Institute of Science and Technology, Vandalur, Chennai, 600048, Tamil Nadu, India.
**Corresponding author: S. Hemalatha,*
Corresponding author mail ID: hemalatha.sls@crescent.education,

ABSTRACT

The green synthesis of nanoparticles offers an eco-friendly, sustainable approach to developing advanced antimicrobial agents. This study reports the synthesis of silver nanoparticles (CtAgNPs) by using *Clitoriaternatea* floral extract via a microfluidic serpentine micromixer. The CtAgNPs were characterized using UV-Visible spectroscopy, dynamic light scattering (DLS), field emission scanning electron microscopy (FE-SEM), X-ray powder diffraction (XRD), energy-dispersive X-ray analysis (EDAX), and Fourier-transform infrared spectroscopy (FTIR), confirming their optimal size, crystallinity, and stability. The antimicrobial activity of CtAgNPs was evaluated in dental pathogens, specifically *Streptococcus mutans* and *Enterococcus faecalis*, key contributors to dental caries. The CtAgNPs effectively inhibited growth and biofilm formation. Oxidative stress was induced in the bacterial cells, as evidenced by altered levels of Glutathione (GSH), Malondialdehyde (MDA), Catalase (CAT), Superoxide dismutase (SOD), and leakage of proteins and sugars, this suggests that nanoparticles disrupted the cellular homeostasis and oxidative system to inhibit the multiplication of the bacterial cells. These findings establish the potential of green-synthesized CtAgNPs as a sustainable solution to combat dental pathogens, providing a promising alternative for the development of innovative antimicrobial agents to control dental caries.

KEYWORDS: *Microfluidics, Clitoriaternatea, Green synthesis, Silver nanoparticles, Dental pathogens, Biofilm inhibition.*

USE OF TANNERY SOLID WASTE-DERIVED MICROBES IN AZO DYE DECOLORIZATION AND DEGRADATION

Aarthi P, Sridevi S, Gokul O, Hemalatha S, Faridha Begum I*

School of Life Sciences, B. S. Abdur Rahman Crescent Institute of Science and Technology, Vandalur, Chennai-600048, Tamil Nadu, INDIA.

**Corresponding author: Faridha Begum I, Corresponding author mail ID: faridhabegum.sls@crescent.education,*

ABSTRACT

Tannery effluent is the primary source of water pollution which has a detrimental impact on both the environment and human health due to the release of azo dye from tannery industry. Reactive dyes, synthetic azo dyes and numerous other hazardous chemicals are among the toxic materials that are released as coloured effluents. These materials alter pH levels and change the organic and inorganic content of water bodies. The research focus on isolating azo dye degrading bacteria from tannery solid waste. A bacterium isolated, screened was employed to decolorize an azo dye and identified as *Thioalbus denitrificans (PP767757), Ectobacillus funiculus (PP767756).* The novel bacterial strain rapidly decolourized the synthetic azo dyes such as black (91%), brown (90%) and red (96%), respectively. The azo dyes degradation was optimized with 10% inoculum concentration at pH7.0, temperature 37°C and 48 hours. *Ectobacillus funiculus* and *Thioalbus denitrificans* can be used as eco-friendly biological expedients for the bioremediation and detoxification of azo dyes using in industries.

KEYWORDS: *Novel bacteria, tannery, bioremediation, degradation, azo dyes.*

HYBRID COMBINATION OF MICROBIAL AND CHEMICAL BIOSENSORS IN ENVIRONMENTAL POLLUTANT DETECTION: UNLOCKING FUTURE PROSPECTS

S. Keerthana

Department of biotechnology, Rajalakshmi Engineering College (REC), Chennai, Tamil Nadu, India.

ABSTRACT

Hybrid biosensors combining microbial and chemical components offer a promising approach for environmental monitoring, providing enhanced sensitivity and selectivity in detecting pollutants. These biosensors integrate the metabolic activity of microorganisms with chemical interfaces or enzymes, enabling accurate identification and quantification of various environmental contaminants. Recent advances have seen the development of nano-enabled hybrids, enzyme-based microbial systems, and whole-cell microbial biosensors integrated with chemical interfaces. Despite the potential, challenges such as labor-intensive preparation, limited research, and scalability issues persist. This review provides an overview of the different hybrid biosensor types, their usage in environmental pollutant detection, highlighting their applications, advantages, and the challenges that need to be addressed for further development. The future prospects for these biosensors in environmental monitoring and pollution control are promising, provided that advances in technology and further research into their integration are made.

KEYWORDS: *Hybrid biosensors, Environmental monitoring, nano-enabled hybrids, enzyme-based microbial system, pollution detection, microbial-chemical integration, biosensor technology.*

INDIAN HERBAL DERIVATIVE FOR LUMBAR PAIN MANAGEMENT IN POST C – SECTION WOMEN: A COMPREHENSIVE REVIEW

K Mythili[1], Dr. S. Prabakar[2], L Swetha[3], AP Saranya[4], G Rajeshwari[5]

Assistant Professor[1], Department of Biomedical Engineering, Sona college of Technology, Salem, Tamil Nadu, India.
Head of the Department[2], Department of Biomedical Engineering, Sona college of Technology, Salem, Tamil Nadu, India.
Final year students[3,4,5], Department of Biomedical Engineering, Sona college of Technology, Salem, Tamil Nadu, India.

ABSTRACT

Many women have distressing lower back pain due to conditions like menstrual cramps, endometriosis, a poor posture, ovarian cysts, menopause and especially after a C-section if their pains persist after giving birth. In this paper, we'll specifically be discussing women who develop Lower Back Pain(LBP) as a result of a Cesarean-Section(C-Section)delivery, but whose pain may turn chronic. Most contributors to this chronic pain are often associated with anesthesia at the time of the procedure, high BMI, a bad posture, hormonal disturbances, heavy lifting, and long-term breastfeeding. Traditional treatments include exercises, heat or cold packs, and medicines that risk causing injury or even side effects such as nausea and dizziness. Such challenges are promisingly addressed by the ethnopharmacological remedies. Moringa oleifera, quite famously known as the "drumstick tree" or the "miracle tree," has been utilized since very olden times due to its several health benefits. Leaves, seeds, and pods of the plant have replete content of vitamins, minerals, and antioxidants, being either anti-inflammatory and antimicrobial. Moringa also controls sugar and cholesterol levels in the blood, aids in digestive health, and promotes general well-being. The mechanism behind Moringa's action in pain management works according to the Gate Control Theory of Pain, thus working as a natural analgesic that modulates inflammatory pain through the suppression of the NF-κB pathway. This will suppress pro-inflammatory cytokines and enzymes, thus bringing about a decrease in local inflammation and pain signals. It therefore stimulates larger nerve fibers, which enhances the closure of the pain gate. With minimal side effects and effective ethnopharmacological actions, Moringa happens to be an ideal treatment option for most women with LBP cases post C-section deliveries. Thus, the objective of this article is to review and summarize the natural herbal derivatives primarily used for LBP which focuses on women health after C-Section.

KEYWORDS: *C-section, Ethnopharmacological remedies, Gate Control Theory of Pain, Inflammation, Lower Back Pain, Moringa oleifera.*

AI-POWERED PREDICTIVE MODELS FOR ENHANCING DRUG-NANOPARTICLE INTERACTIONS IN NANOMEDICINE

S. Sona

Department of Bioinfomatics, Saveetha School of Engineering, Saveetha University, Saveetha Nagar, Sriperumbadur Taluk, Kanchipuram - Chennai Rd, Chennai, Tamil Nadu 602105

ABSTRACT

The utilization of nanoparticles in drug delivery systems represents a significant leap in modern medical treatments, offering improved bioavailability, stability, and targeted drug release. Despite their immense potential, accurately predicting the interactions between nanoparticles and drugs remains a formidable challenge due to the inherent complexity of their behaviors in biological environments. Traditional experimental methods for studying these interactions are often cost-prohibitive and time-consuming. In this research, we propose an innovative approach using artificial intelligence (AI) to develop predictive models that can accurately simulate and optimize drug-nanoparticle interactions. Through the application of machine learning algorithms, we analyze key physicochemical characteristics of nanoparticles and drugs, along with essential biological factors, to construct highly reliable models. The findings of this study underscore the efficacy of AI-based approaches in reducing time and costs while enhancing the precision of nanoparticle drug delivery systems. The results also highlight critical interaction parameters such as nanoparticle size, surface charge, and the molecular properties of the drug, paving the way for future improvements in nanomedicine.

KEYWORDS: *Nanoparticles, Drug Delivery, Artificial Intelligence, Machine Learning, Algorithms.*

CO_2 VALORIZATION INTO USEFUL ENZYMES USING MICROBIAL ELECTRO SYNTHESIS

C Gayathri, P Sagana

UG Students, Department of Biotechnology, Sree Sastha Institute of Engineering and Technology, Chembarrambakkam, Chennai-600123.

ABSTRACT

Climate change mitigation strategies focus on reducing CO2 emissions and developing sustainable technologies for CO2 valorization. Microbial electrosynthesis (MES) is an innovative approach that utilizes microorganisms to convert CO2 into valuable chemicals. Previous studies have explored the potential of *Saccharomyces cerevisiae* for MES-based production of lycopene, a high-value carotenoid pigment. It's a novel Bioelectrochemical system (BES) that integrates *S. cerevisiae* with an electrochemical cell. The BES was operated under controlled conditions, and the effects of various parameters, including electrode potential, CO_2 flow rate, and nutrient supply, were investigated. Previous results have also demonstrated that *S. cerevisiae* can efficiently convert CO_2 into lycopene through MES. Thus, MES-based CO_2 valorization using *S. cerevisiae*can be used for the production of high-value chemicals.

KEYWORDS: *CO_2 Valorization, Microbial electrosynthesis, Saccharomyces cerevisiae, Lycopene, Bioelectrochemical systems, HPTLC.*

ROLE OF DARK OXYGEN IN MARINE LIFE: INSIGHTS

A. Vishal,C. R. Kanishkar, D.Venkatesan

UG Students, Department of Biotechnology, Sree Sastha Institute of Engineering and Technology, Chembarrambakkam, Chennai-600123.

ABSTRACT

Dark oxygen mainly originates from polymetallic nodules found in the CCZ. These are coal-like mineral rocks which contains manganese and iron. Unlike the oxygen produced by photosynthesis from plants and algae, these nodules generate oxygen without sunlight. However, at depths exceeding 5km where sunlight cannot reach, these natural metal formations appear to catalyze the splitting of seawater (H2O) into hydrogen and oxygen, which contributes significant oxygen supply for oceans. In deep-sea environments devoid of sunlight and plants, some life forms get their energy from chemicals that erupt from the sea floor at hydrothermal vents. A recent research by *Sweetman et al. 2024* on the production of dark oxygen by polymetallic nodules showed that it not only reveal unexpected truths, but also highlight the importance of acting according to the Precautionary Principle.

KEYWORDS: *Polymetallic nodules, Electrolysis, Survival, Aquatic photosynthesis.*

PRODUCTION OF MORPHINE FROM PAPAVER SOMNIFERUM: ANTIMICROBIAL APPLICATIONS

D. Shalini, I. Blenda Patricia

UG Students, Department of Biotechnology, Sree Sastha Institute of Engineering and Technology, Chembarrambakkam, Chennai-600123.

ABSTRACT

Papaver somniferum (opium poppy) is a main source of several pharmaceutical benzylisoquinoline alkaloids such as morphine, codeine and sanguinarine. *Papaver* compounds display antioxidant, antimutagenic, and anticarcinogenic effects, antimicrobial and antiparasitic activities, anti-inflammatory activity, and its effect on nervous system, and also other effects. The extraction of pharmaceutically important alkaloids like morphine and codeine from poppy plant meets the need for the most suitable and standard methods in pharmaceutical industry. Of the opium alkaloids found on poppy seeds, morphine, and codeine are the most pharmacologically active and have been detected in biological matrices collected in workplace and roadside drug testing resulting in positive opiate results. In recent research, maximum concentrations of 530, 60, and 120 mg kg^{-1} for morphine, codeine, and the baine were reported respectively.

KEYWORDS: *Opioid, Pharmacological potency, Biosynthetic pathway, Antimicrobials.*

PRODUCTION OF POLYHYDROXYBUTRATE USING NATURAL AGRO WASTE: GROUNDNUT SHELL

Puratchikavi V, Sandhiya S, Sowmiya

UG Students, Department of Biotechnology, Sree Sastha Institute of Engineering and Technology, Chembarrambakkam, Chennai-600123.

ABSTRACT

Polyhydroxybutyrate (PHB) is a monomer of polyhydroxyalkanoates (PHA) that is biosynthesized, accumulated in intracellular lipid granules, and are mainly produced by plenty of microorganisms. PHB biopolymer has numerous applications in medicine, agriculture, automobiles, packaging, and cosmetic industries. *Bacillus* sp., *Azotobacter* sp., *Cupriavidus* sp., and *Pseudomonas* sp. were among the most dynamic microbes that produce PHB biopolymer. Groundnut shell is considered to agro-industrial waste product and it is rich in lignocellulose materials. The biopolymer polyhydroxybutyrate (PHB) and biopigment carotenoid produced by microbes holds significant industrial and environmental potential. According to the recent survey, it furnishes an efficient agricultural waste executive approach by diminishing the production expenditure. It reinforces the production of PHB, thereby shrinking our reliance on fossil fuel-based plastics.

KEYWORDS: *Polyhydroxybutyrate (PHB), Agro-waste, Biopolymer, Ecofriendly.*

HOLY BASIL LEAF NANOPARTICLES COATING FOR PROLONGING FOOD SHELF LIFE

E. Bharathi, S. Suguna, U. Abirami

K.S.Rangasamy College of Technology, KSR Kalvi Nagar, Tiruchengode,Namakkal (DT),637215.

ABSTRACT

The preservation of perishable vegetables, such as green chilli and okara (Abelmoschus esculentus), is a major concern due to microbial spoilage and loss of nutritional value. This study focuses on the eco-friendly synthesis of silver nanoparticles (AgNPs) using Holy Basil (Ocimum sanctum) leaves, leveraging their phytochemical properties for applications in food coatings and packaging. The synthesis of AgNPs is facilitated using basil leaf extract as a natural reducing and stabilizing agent. Phytochemical analysis of the extract reveals the presence of bioactive compounds, including flavonoids, phenolics, and terpenoids, which play a critical role in nanoparticle formation and stabilization. The synthesized nanoparticles are characterized through UV-Vis spectroscopy, Transmission Electron Microscopy (TEM), scanning electron microscopy (SEM), and dynamic light scattering (DLS) to determine their structure, size, and stability. Antimicrobial and antioxidant assays demonstrate the efficacy of AgNPs in inhibiting the growth of foodborne pathogens such as Escherichia coli and *Aspergillus niger* while preventing oxidative degradation of nutrients in the vegetables. The nanoparticles are then integrated into biopolymer-based edible coatings and packaging materials, tested for their application on green chilli and ladies' finger. Shelf-life studies reveal that the coating materials significantly reduce microbial spoilage, delay dehydration, and preserve the nutritional and sensory qualities of the vegetables over extended storage periods. These findings underscore the potential of Holy Basil-derived AgNPs as an innovative and sustainable solution for food preservation, reducing post-harvest losses and promoting ecofriendly practices in the food industry.

KEYWORDS: *Preservation, Holy Basil, Abelmoschus esculentus, Green chilli, Silver Nanoparticles.*

ISOLATION AND GENOME SEQUENCING OF NOVEL RHIZOSPHERE BACTERIAL STRAIN RESPONSIBLE FOR ENHANCED CROP YIELD

Hemavathi B, Arun Arumugaperumal

Department of Biotechnology, Rajalakshmi Engineering College, Thandalam, Chennai-602105.

ABSTRACT

The rhizosphere, a critical soil region surrounding plant roots, harbors a diverse and dynamic community of microorganisms that profoundly influence plant growth and crop productivity. Among these, rhizosphere-associated bacteria play pivotal roles in nitrogen fixation, phosphate solubilization, phytohormone production, and pathogen suppression. This study investigates the isolation, identification, and functional characterization of beneficial rhizobacterial strains that promote crop health and enhance yield. Advances in molecular approaches, including genomics and metagenomics, have enabled the discovery of novel bacterial strains with superior plant-growth-promoting traits. Additionally, this research explores the challenges of implementing these strains in large-scale agricultural systems and evaluates their potential in fostering sustainable farming practices. Understanding the intricate interactions between plants and beneficial rhizobacteria offers a pathway to developing eco-friendly, bio-based solutions for improving crop productivity and soil health.

KEYWORDS: *Rhizosphere, Rhizobacteria, Sustainable Agriculture, Whole Genome Sequencing.*

WHOLE GENOME SEQUENCING OF BACTERIA FROM NEPHRIDIA OF EARTHWORM

Chandhana U, Chandhana NM, Arun Arumugaperumal

Department of Biotechnology, Rajalakshmi Engineering College, Thandalam, Chennai-602105.

ABSTRACT

Earthworms serve as vital ecosystem engineers, significantly shaping soil structure while enhancing nutrient cycling and the breakdown of organic matter. Their nephridia, which function similarly to vertebrate kidneys, create a specialized environment that supports bacterial colonization. Through whole-genome sequencing, this study thoroughly examines the bacterial communities residing within earthworm nephridia. We are confident that these bacteria play a crucial role in various aspects of earthworm physiology, including detoxification, nutrient processing, and immune modulation. The bacteria released into the soil via earthworm excreta are likely to further influence essential soil processes.

KEYWORDS: *Earthworm, Nephridia, Symbiote, Whole genome sequencing.*

ABILITY OF *LACTOBACILLUS CASEI* IN PURE COW MILK TO ABSORB HEAVY METALS BY ICP-MS: A COMPARISON WITH *LACTOBACILLUS FERMENTUM*

Y. Jahnavi, Megna, G D. Yogavarshan, K. Nithyalatha

SIMATS Engineering, Saveetha Nagar, Thandalam,Chennai-602105.

ABSTRACT

Heavy metal contamination, particularly cadmium, in food products is a significant environmental and public health concern due to its bioaccumulation and toxicity. Probiotic bacteria such as *Lactobacillus casei* and *Lactobacillus fermentum* are being explored as natural detoxification agents for heavy metals in food systems. This study investigates the cadmium adsorption efficiency of *Lactobacillus casei* in pure cow milk and compares it with *Lactobacillus fermentum*. Pure cow milk samples were inoculated with each probiotic strain, incubated for 24 hours, and analyzed for cadmium removal using Inductively Coupled Plasma Mass Spectrometry (ICP-MS). The results show that *L. fermentum* achieved a cadmium removal rate of 68.3%, slightly higher than *L. casei* at 63.5% ($p < 0.05$).The adsorption mechanism is attributed to extracellular polysaccharides on the bacterial surface, facilitating ion binding. This study highlights the potential of *Lactobacillus casei* as a natural adsorbent for heavy metals, offering a sustainable solution for improving milk safety and reducing health risks associated with heavy metal exposure.

KEYWORDS: *Lactobacillus casei, Lactobacillus fermentum, Bioaccumulation, Toxicity.*

HALOPHYTE-ASSOCIATED PGPR AND ENDOPHYTES: A SUSTAINABLE SOLUTION FOR SALT-STRESSED CROPS

Joseph Ezra John[1], Chidamparam Poornachandhra[2*], Boopathi Gopalakrishnan[3], Perumal Thangavel[2]

[1] *Department of Environment and climate change, Govt. of Tamil Nadu, Chennai, India*
[2] *Department of Environmental Sciences, AC&RI, TNAU, Coimbatore, India*
[3] *ICAR-National Institute of Abiotic Stress Management, Baramati, India*
* *Correspondence:poorna155c@gmail.com*

ABSTRACT

Soil salinity adversely impacts agricultural productivity and has become a widespread issue globally, intensifying over recent decades due to excessive fertilizer use, poor-quality irrigation water, and wastewater reuse, which result in salt accumulation in the soil. Halophytes, however, thrive in highly saline environments, primarily due to the beneficial bacterial microbiome associated with their roots that enhances salt tolerance. In this study, salt-tolerant bacteria with multiple plant growth-promoting (PGP) characteristics were isolated from the rhizosphere of the halophyte *Sesuvium portulacastrum*, cultivated in coastal and paper mill effluent-irrigated soils with electrical conductivity (EC) exceeding 3.0 dS m^{-1}. Among the isolates, nine elite halotolerant rhizobacterial strains were identified, including *Metabacillusindicus*, *Neobacillusniacini*, *Staphylococcus warneri*, *Bacillus velezensis*, *Bacillus circulans*, *Kocuriarhizophila*, *Bacillus oleronius*, *Serratiamarcescens*, and *Kosakoniaradicincitans*. These strains exhibited robust growth at salinity levels up to 5% NaCl and demonstrated significant PGP traits, notably 1-aminocyclopropane-1-carboxylic acid (ACC) deaminase activity and indole acetic acid (IAA) production, which mitigate stress ethylene in plants under salinity. To evaluate their efficacy in enhancing salt tolerance in glycophytes, *Vignamungo* L. seeds were inoculated with these strains and subjected to induced salinity conditions (2% NaCl). The inoculated seeds showed significant improvements in germination rate, shoot length, and vigor index compared to non-inoculated seeds under saline conditions. Among the tested strains, seeds inoculated with *S. marcescens* achieved the highest germination rate of 89%, followed by *B. velezensis* at 85%. Additionally, seeds inoculated with *B. velezensis* exhibited a Vigor index twofold higher than that of non-inoculated seeds under 2% NaCl. The findings suggest that halotolerant PGPR isolated from *S. portulacastrum* offer a cost-effective and environmentally sustainable approach to enhance the productivity of *Vigna mungo* L. in saline conditions.

KEYWORDS: *Halophyte; Rhizobacteria; Endophytes; Salt tolerance; Crop improvement*

BIOACTIVE METABOLITES ISOLATED FROM ENDOPHYTIC FUNGUS FROM THE LEAVES OF *SYZYGIUM CUMINI* FOR TREATING BACTERIAL DISEASES IN AQUACULTURE

Sujithra Vijayakumar, John Thomas*

**Corresponding author: Dr. John Thomas, Associate Professor, Centre for Nanobiotechnology, Vellore Institute of Technology (VIT), Vellore, Tamil Nadu, India.*
Corresponding author mail ID: john.thomas@vit.ac.in

ABSTRACT

The infections caused through the bacterial pathogens is a significant concern affecting the aquaculture sector. The present study aims to isolate a natural colorant source with potential bioactive compound for application in the aquaculture. In the current study, a fungal endophytic isolate (denoted as VITJT) was isolated from the leaves of Syzygium cumini. The isolate VITJT was characterized through molecular and phylogenetic tree analysis for the identification of the isolate. The secondary metabolites were extracted from the fungal isolate. The crude secondary metabolites from the fungal isolate VITJT were analyzed through Ultraviolet-Visible spectroscopy, Fourier transform infrared spectroscopy and GCMS to identify the chemical components present in the extract. The GC-MS analysis revealed the presence of bioactive compounds in the extract. The toxicity (LC50) of the crude secondary metabolites were analyzed through brine shrimp lethality assay. The LC50 was found to be above 200 μg/mL in the brine shrimps. The secondary metabolites from the fungal isolate possessed antibacterial activity against the aquatic pathogens, *Aeromonas hydrophila* and *Vibrio parahaemolyticus*. A significant zone of inhibition of20.33±0.33 mm was observed in the maximum concentration of 200 μg/mL againstthe *A. hydrophila*. In the case of *V. parahaemolyticus*, an inhibitory zone of 20±0.28mm was observed at the maximum concentration of 200 μg/mL. The secondary metabolites possessed antioxidant activities as well. The DPPH scavenging activity for the crude metabolites at the maximum concentration (100 μg/mL) was identified as 68.57 ±0.16 (%). The hydrogen peroxide assay for the crude metabolites at the maximum concentration (100 μg/mL) was identified as 56.25 ±0.56 (%). The crude secondary metabolites are a natural source which can be applied as a natural colorant with antibacterial and antioxidant properties.

KEYWORDS: *Endophytic fungus; Syzygium cumini; Aquatic pathogens, Antioxidant; Antibacterial.*

ESTIMATION OF REACTIVE OXYGEN SPECIES AND STRESS-INDUCED APOPTOSIS LEVELS IN THE ZEBRAFISH MODEL EXPOSED TO CATLA FISH-DERIVED HEPCIDIN PEPTIDE AND HYDROGEN PEROXIDE USING FLUORESCENT STAINING FOR ENHANCED ANTIOXIDANT PROPERTY.

B.Subba Deekshitha, Y.Yogitha Reddy, Tanikonda Venkata Divya

Department of Biotechnology, Saveetha School of Engineering, Thandalam, Chennai-602105.

ABSTRACT

Reactive oxygen species (ROS) and oxidative stress are major contributors to cellular damage, leading to apoptosis and various oxidative stress-related disorders. Identifying effective antioxidants is crucial for mitigating these effects. This study investigates the antioxidant and cytoprotective potential of *Catla catla*-derived hepcidin peptide in a zebrafish model exposed to hydrogen peroxide (H_2O_2)-induced oxidative stress. Zebrafish embryos were treated with H_2O_2 to induce ROS generation and apoptosis, followed by exposure to hepcidin peptide. ROS levels were quantified using DCFH-DA fluorescent staining, while apoptosis was assessed with acridine orange staining. Fluorescence microscopy was used to visualize and measure the intensity of ROS and apoptotic signals.The results showed that embryos treated with hepcidin peptide exhibited a significant reduction in ROS levels compared to the H_2O_2-only group, indicating the peptide's strong antioxidant activity. Furthermore, acridine orange staining revealed a marked decrease in apoptosis in the peptide-treated group, suggesting enhanced cellular protection against oxidative damage. These findings demonstrate that Catla hepcidin peptide effectively mitigates ROS-induced oxidative stress and prevents stress-induced apoptosis, likely through its antioxidant and cytoprotective mechanisms. This study highlights the potential of Catla hepcidin peptide as a natural therapeutic agent for combating oxidative stress-related cellular damage. Its ability to reduce ROS levels and apoptosis underscores its promise in developing novel treatments for oxidative stress-associated disorders. Further research could explore its molecular mechanisms and therapeutic applications in clinical settings.

KEYWORDS: *Reactive oxygen species, Oxidative stress Catla catla, Zebrafish.*

EXPLORING THE IMPACT OF PHYTO COMPATIBILITY SCREENING ON CELL APOPTOSIS

Yannabathula Vaishnavi, K. Poojitha, Nishicaa. P, P, Jeevitha

Department of Biotechnology, Saveetha School of Engineering, Thandalam, Chennai-602105.

ABSTRACT

Phyto-compatibility screening of plant-derived compounds is a crucial step in identifying therapeutic agents with minimal toxicity and maximum efficacy. This study explores the relationship between phyto-compatibility and the regulation of cell apoptosis, a fundamental process in maintaining cellular homeostasis and targeting cancer cells. By evaluating a diverse range of phytochemicals, we aim to assess their ability to modulate apoptosis pathways, including intrinsic and extrinsic signalling mechanisms. The experimental approach includes high-throughput screening of plant extracts for cytotoxicity, their effect on mitochondrial membrane potential, caspase activation, and pro- and anti-apoptotic protein expression. Our results reveal that several phytochemicals exhibit selective cytotoxicity towards cancer cells while sparing normal cells, highlighting their potential as apoptosis-inducing agents. Notably, compounds exhibiting optimal phyto-compatibility were shown to synergize with conventional therapies, enhancing apoptosis and reducing resistance. This study underscores the importance of phyto-compatibility screening in the discovery of plant-based therapeutics for apoptosis regulation. The findings pave the way for developing novel, less toxic, and more effective treatments for cancer and other diseases involving dysregulated apoptosis.

KEYWORDS: *Phyto-compatibility, Cancer, Apoptosis, Caspase, Cytotoxicity.*

DEVELOPING A BIODEGRADABLE TABLET COATING TECHNOLOGY THAT ACHIEVES CONTROLLED DRUG RELEASE WHILE MINIMIZING ENVIRONMENTAL IMPACT

Darsi. Asritha, Nimmakayala Divyanjali, Muramreddy Sreeja

Saveetha School of Engineering, SIMATS, Thandalam, Chennai-602105.

ABSTRACT

By minimizing ecological harm with effective therapeutic outcomes, biodegradable tablet coating development has attracted interest due to increasing environmental concerns regarding pharmaceutical packaging and drug delivery systems. This research seeks to develop a new biodegradable tablet coating technology which is both capable of controlled drug release, and able to satisfy the growing demand for sustainability in the pharmaceutical industry. Using biopolymers like cellulose derivatives, starch and chitosan, the present study investigated green materials for coatings that could help to control the release of active pharmaceutical ingredients (APIs). Our coatings are biodegradable under specific physiological conditions, making them environmentally friendly and stable at storage and transportation. Controlled dimensions and property features make the thin sheet film potential drug carrier for a sustained and targeted delivery of APIs which is possible through controlled drug release implemented by careful customization of coating thickness, composition, and surface properties. It also aims to study the interaction of coating material with drugs, and that of degradation kinetics, in order to obtain an ideal release profile. The ultimate goal is to design a technology that combines adequate pharmaceutical function of the tablets grade with the degradability properties minimizing dependence on synthetic non-degradable polymers. This has implications for sustainable pharmaceutical ways, plastic waste reduction and could help impact patient outcomes through effective drug delivery systems.

KEYWORDS: *Biodegradable tablet, Coating, Pharmaceutical ingredients, Plastic waste, reduction.*

DETERMINING MARINE SEDIMENT SAMPLES FOR MELANIN-PRODUCING BACTERIA

Murshidah Shereen.M, Noble K Kurian,

School of Life Sciences, B.S. Abdur Rahman Crescent Institute of Science and Technology, Chennai, India.Author mail id: murshidahshereen26@gmail.com

ABSTRACT

The colour of bacterial melanin is either brown or black. Under stress, bacteria create melanin, which serves as a shield to protect them from pathogens, heavy metals, UV radiation, and other harmful substances. Melanin exhibits excellent optical absorbance, photoprotective, and antioxidant properties. The purpose of this study is to isolate melanin-producing bacteria from a marine sediment sample from Thiruvanmiyur beach, Chennai India. Bacteria producing melanin were isolated from the sediment sample. Strains CTB 3, CTB 5, CTB 7, and CTB 10 was found to produce good amount of melanin. The extracted melanin sample will be subjected to characterization (UV-visible spectroscopy) and chemical analysis. Compared to the other three strains, the CTB 10 strain produces melanin faster. This study advances the creation of an environmentally safe and sustainable melanin production process that may be used in sunscreen, to replace synthetic dyes and absorb heavy metals from the water.

KEYWORDS: *Melanin, Stress protection, Bacteria, Marine, Photo protection.*

BIOREMEDIATION OF PAINT INDUSTRY EFFLUENT USING NOVEL BACTERIAL STRAINS

Ajay Rathnam B, Faridha Begum I*

School of Life Sciences, B. S. Abdur Rahman Crescent Institute of Science and Technology, Vandalur, Chennai-600048, Tamil Nadu, INDIA.
**Corresponding author: Faridha Begum I, Corresponding author mail ID: faridhabegum.sls@crescent.education,*

ABSTRACT

Paint industry effluent contains a complex mixture of hazardous chemicals, including heavy metals, solvent and synthetic polymers, which pose synthetic environmental challenges. This study focus on isolating, identifying novel bacterial strains capable of degrading paint industry effluent and producing biosurfactant. Four bacterial strains where identified through 16S rRNA sequencing including *Rossellomorea marisflavi, Sutclifella zhanjiangensis, Jeotgalibacillus aurantiacus* and *Kosakonia sacchari.* Among these *Rossellomorea marisflavi* exhibited highest degradation efficiency of 71.63%. Biosurfactant production was confirmed using assays such as drop collapse, oil spread, emulsification, BATH and Biuret test, with significant activities demonstrated by the isolates. Future work involves optimization of degradation conditions and detailed chemical characterization of biosurfactants using advanced analytical techniques. These findings highlight the potential of these bacterial strains for sustainable bioremediation and environmental management strategies.

KEYWORDS: Paint industry effluent, Bioremediation, Biosurfactant production, Sustainable solution.

EXPLORING T-CELL IMMUNE EVASION BY LEISHMANIA: ANALYSING THE ROLE OF IMMUNE MODULATION IN *VISCERAL LEISHMANIASIS* PATHOGENESIS

S. Tanushree, A.Priyanka, P. Mathu Sri

Saveetha Institute of Medical and Technical Sciences SIMATS, Chennai.

ABSTRACT

Visceral leishmaniasis (VL), a life-threatening disease caused by *Leishmania donovani*, is a significant public health concern, particularly in resource-limited regions. The parasite's ability to evade host immune defenses, especially T-cell-mediated immunity, is central to its pathogenesis. While previous studies have identified some immune evasion strategies of *Leishmania*, there is limited understanding of how the parasite systematically modulates T-cell responses to establish chronic infection and disease progression. This study aims to explore and analyze the mechanisms by which *Leishmania* modulates T-cell activity, focusing on its role in immune evasion and the progression of *Visceral leishmaniasis*. Additionally, the study will examine the therapeutic implications of these immune modulation strategies, specifically how targeting these processes could enhance the immune response and lead to more effective treatments for VL. The research is based on an extensive review of existing literature, including research articles, reviews, and case studies. Key topics explored include cytokine modulation, the induction of regulatory T cells (Tregs), alterations in antigen presentation, and immune checkpoint molecule activation. By understanding the immune modulation strategies employed by *Leishmania*, this research highlights potential targets for therapeutic intervention. Restoring effective T-cell responses through immunotherapies or vaccines could significantly improve outcomes for patients with visceral leishmaniasis.

KEYWORDS: *Visceral leishmaniasis, Leishmania donovani, cytokine modulation, immunotherapies, vaccines.*

UNVIELING THE POTENTIAL OF CORN SILK AS DIURETIC AGENT IN DIABETIC NEPHROPATHY

Priya D, Kalpana K, T.R. Poorani

Department of Biotechnology, Rajalakshmi Engineering College, Chennai.

ABSTRACT

Diabetic nephropathy (DN) is the most common and serious microvascular complication in diabetic patients, and it easily develops into end-stage renal disease compared to other causes. The clinical treatment of DN mostly adopts early intervention, which mainly focuses on controlling blood sugar and blood pressure first and then protecting renal function. Although this can delay the progress of DN, it cannot completely stop its progress. DN is closely related to abnormal glucose metabolism, hemodynamics, the renin angiotensin aldosterone system, advanced glycation end products, inflammation and oxidative stress. The gut microbiota is a complex ecosystem that has been described as the "second genome" controlling human health. Polysaccharides can significantly reduce chronic low-grade inflammation by balancing gut microbiota, leading to improvements in metabolic disorders such as blood glucose, obesity, and lipid metabolism. The phytochemical components present in corn silk-like flavonoids, phenolics, terpenoids, tannins, sterols, and alkaloids are phytochemical components that have hypoglycemic activity and a mechanism for lowering blood glucose levels. It is a diuretic that can promote diuresis and reduce swelling, soothe the liver and promote bile flow, eliminate dampness and alleviate jaundice. corn silk as a medicine, which can be used for acute and chronic nephritis, edema, proteinuria, and diabetes. This suggests that corn silk may modulate key metabolites associated with end-stage renal function failure, and these metabolites are closely linked to the gut microbiota. DN rat model, and introduced the concept of the gut-kidney axis.

KEYWORDS: *Diabetic nephropathy, Corn silk, Polysaccharides.*

ADVANCED BIOPROCESSING OF LEATHER SOLID WASTE

Selina M, Faridha Begum I*

School of Life Sciences, B.S Abdur Rahman Crescent Institute of Science and Technology, Vandalur, Tamil Nadu India.

**Corresponding author: Faridha Begum I, Corresponding author mail ID: faridhabegum.sls@crescent.education*

ABSTRACT

The leather industry is one of the industries that produces significant amounts of solid waste, such as shavings, fleshing, and trimming, as a result of its various processes. These solid wastes have high impact on environment by leeching out into the soil and are difficult to dispose of. Effective degradation using advanced bioprocessing helps overcome this situation and supports environmental sustainability. This study explores into innovative biotechnological approach to enhance waste management by microbial fermentation and green chemistry. Microbial fermentation was employed as a key approach to degrade the leather solid waste using specific microbial consortia capable of breaking down the collagen-rich skin. Specific microbial strains showed high efficiency in degrading the leather waste into precursors for biofertilizers and biochemicals, ensuring an economically viable way for creating high-value products. This study illustrates the viability of a circular economy model for the leather industry by incorporating cutting-edge bioprocesses, especially microbial fermentation, and offers a road map for value generation and sustainable waste management. This is a 'Waste to Wealth mission' which focuses on identifying, developing, and implementing technologies fir recycling materials and extracting valuable resources from waste, not only minimises the environmental burden but also creates new economic opportunities and contributes to Sustainable Development Goals (SDGs).

KEYWORDS: *Waste to Wealth mission, Sustainable Development Goals (SDGs), Recycling technologies, Waste valorization, High-value products.*

BIOCOMPATIBLE SILVER NANOPARTICLES TO DEVELOP DERMO-COSMETIC AND DERMATOLOGICAL PRODUCTS FOR CELLULITIS

S. Jaswanthika, Ranjani S, S. Hemalatha

School of Life Sciences, B. S. Abdur Rahman Crescent Institute of Science and Technology, Vandalur, Chennai, 600048, Tamil Nadu, India.
**Corresponding author: S. Hemalatha,*
Corresponding author mail ID: hemalatha.sls@crescent.education,

ABSTRACT

Our skin is the first line of defense against various infections. A systemic response is activated when this line of defense is disturbed. Skin infection could be caused by bacteria either invading normal skin or affecting a compromised skin barrier. Cellulitis is a common acute infection of the skin and subcutaneous tissue. People affected by this disease will have a burden in secondary care due to infections. The common signs of cellulitis are rubor, dolor, tumor and calor. The cellulitis caused by multi-drug-resistant *Escherichia coli* and *Klebsiella pneumoniae* is mainly focused in this research work. Silver nanoparticles are outstanding antimicrobial agents, effective against both Gram-negative and Gram-positive bacteria, including those resistant to multiple drugs. This makes them a referable choice to break antibiotic resistance. Silver nanoparticles are a valuable resource for controlling the infections caused by multi-drug resistant organisms. Green synthesis of silver nanoparticles will be explored to control cellulitis causing multi-drug resistant pathogens. The development of novel topical products will pave a new profound way and help in the battling cellulitis.

KEYWORDS: *Skin infection, Cellulitis, Silver nanoparticles.*

GREEN ECO-FRIENDLY NANO PACKAGING MATERIAL TO PREVENT FOOD SPOILAGE

Disha Laljith, Ranjani S and S. Hemalatha

School of Life Sciences, B. S. Abdur Rahman Crescent Institute of Science and Technology, Vandalur, Chennai, 600048, Tamil Nadu, India.
**Corresponding author: S. Hemalatha,*
Corresponding author mail ID: hemalatha.sls@crescent.education,

ABSTRACT

Different types of combinations of organic, inorganic nanoparticles are used in food packaging to prevent losses of food products and to increase the shelf life of food products. Selenium nanoparticles (SeNPs) have attracted much interest due to their remarkable antioxidant qualities, biocompatibility, and capacity to prolong shelf life. This study explores the green synthesis of SeNPs using plant-based reducing agents, offering a sustainable and eco-friendly alternative to conventional methods. Plants, rich in diverse phytochemicals, provide an abundant and renewable resource for nanoparticle synthesis. Incorporating SeNPs into biodegradable food wrappers addresses two critical challenges: enhancing food preservation by mitigating oxidation, moisture absorption, and microbial growth, and reducing environmental pollution through decreased reliance on plastic packaging. SeNP-infused food wrappers not only preserve the nutritional value and flavour of food but also align with the growing demand for sustainable and biodegradable packaging solutions. By integrating nanotechnology with green chemistry, this research highlights the transformative potential of SeNPs in advancing food preservation technologies while minimizing ecological impacts.

KEYWORDS: *Selenium Nanoparticles, Green Synthesis, Biodegradable Food Wrappers, Antioxidant Property, Food Preservation.*

MAGNETIC CORE UREASE PARTICLES FOR UREA REMOVAL IN DIALYSIS SYSTEMS

T.Padmaswarobeeni

Saveetha School of Engineering, SIMATS, Thandalam, Chennai-602105.

ABSTRACT

Chronic kidney disease, a debilitating condition characterized by the gradual deterioration of the kidney's ability to filter waste products from the bloodstream, has emerged as global health concern of significant magnitude, impacting the lives of millions worldwide. A critical component of managing this disease is the process of dialysis, which serves as a vital substitute for the kidney's filtration function. However, the current dialysis methods face several challenges, including the incomplete removal of toxins and middle molecules, as well as the loss of valuable blood components. The development to this innovative technology involves a multifaœted process that begins with the careful selection of materials, drawing on the insights from studies on membrane separation technology and haemodialysis. Researchers have explored the use of magnetic-polymer sphères to selectively bind and remove toxins from the bloodstream, a concept that could be adapted for the removal of urea in dialysis systems.

KEYWORDS: *Urease, Fe_3O_4, Silica Cell Coating, Magnetic Separation Unit.*

ARACHIS HYPOGAEA (NUTSHELL EXTRACT) MEDIATED-GREEN SYNTHESIS OF SILVER AND ZINC OXIDE NANOPARTICLES AND ITS POTENTIAL APPLICATION IN CROP PROTECTION

Vishnu Vardhan G , Ranjani S , Hemalatha S

School of Life Sciences, B. S. Abdur Rahman Crescent Institute of Science and Technology, Vandalur, Chennai, 600048, Tamil Nadu, India.
**Corresponding author: S. Hemalatha,*
Corresponding author mail ID: hemalatha.sls@crescent.education

ABSTRACT:

The growing awareness of the harmful effects of chemical fungicides resulted in the increasing demand for natural and sustainable alternatives for treatment of fungal diseases infecting plants. This study focuses on the green biosynthesis of silver (AgNPs) and zinc oxide nanoparticles (ZnNPs) using the aqueous extract from the nutshell of *Arachis hypogaea*, a common bio-waste. The nutshell extract was prepared and the green nanoparticles were synthesized. The synthesized green silver and zinc oxide nanoparticles were characterized by using UV-Vis spectroscopy, Dynamic Light Scattering (DLS), Zeta Potential analysis, Fourier Transform Infrared (FTIR) spectroscopy, Field Emission Scanning Electron Microscopy (FESEM) with Energy Dispersive X-ray (EDAX) analysis, and X-ray Diffraction (XRD). The Antifungal activity was analysed by assessing the synthesized nanoparticles against four strains of *Aspergillus niger* and the minimum inhibitory concentration (MIC) and minimum fungicidal concentration (MFC) were determined. The results support the potential antifungal activity of *Arachis hypogaea* mediated-green nanoparticles and encourages the development of novel nano-spray formulations to control fungal pathogens infecting plants.

KEYWORDS: *Arachis hypogaea*, Silver nanoparticles, Zinc oxide nanoparticles, Green synthesis, Antifungal Activity, *Aspergillus niger*, Nano-spray.

SYNTHESIS, CHARACTERIZATION AND EXPLORING THE ANTIFUNGAL ACTIVITY OF GREEN NANOEMULSIONS IN CANDIDA SP.

Abdul Muezz K M S, Ranjani S, S. Hemalatha*

School of Life Sciences, B. S. Abdur Rahman Crescent Institute of Science and Technology, Vandalur, Chennai, 600048, Tamil Nadu, India.

**Corresponding author: S. Hemalatha,*

Corresponding author mail ID: hemalatha.sls@crescent.education

ABSTRACT

Nanotechnology has emerged as a promising field in the development of skincare formulations. The demand for natural and safe production of skincare and pharmaceutical products has driven interest in plant-based synthesis of nanoparticles. This study aims to explore the green synthesis of silver nanoparticles (AgNPs) by using an aqueous extract of Citrus maxima. The synthesized AgNPs will be characterized through techniques including UV-Vis spectroscopy, Dynamic light scattering (DLS), Zeta potential analysis, FTIR spectroscopy, FESEM with XRD. These nanoparticles will be incorporated into a nanoemulsion formulation containing lemon oil to enhance its antifungal properties. Antifungal activity of formulated nanoemulsion will be carried out in Candida sp. This study aims to create a promising plant-derived approach that combines the benefits of phytcompounds and silver ions in a stable delivery system.

KEYWORDS: *Silver nanoparticles, Nanoemulsion, Antifungal activity; Candida species, Citrus maxima.*

SYNTHESIS AND CHARACTERIZATION OF NANOPARTICLES TO COMBAT BIOFILM CAUSING FOOD BORNE PATHOGENS

Nivetha J, Ranjani S and S. Hemalatha *

School of Life Sciences, B. S. Abdur Rahman Crescent Institute of Science and Technology, Vandalur, Chennai, 600048, Tamil Nadu, India.
**Corresponding author: S. Hemalatha,*
Corresponding author mail ID: hemalatha.sls@crescent.education

ABSTRACT

The aqueous extracts of *Eclipta prostrata*, *Phyllanthus amarus*, and *Acalypha indica* were utilized for the eco-friendly green synthesis of silver nanoparticles. These plants, known for their antioxidant, antibacterial, and anti-inflammatory properties, contain bioactive compounds including flavonoids, phyllanthin, hypophyllanthin, and alkaloids, which contribute to enhanced nanoparticles bioactivity through a synergistic effect. The characterization of synthesized nanoparticles was performed by using UV-Vis spectroscopy, FTIR, EDAX, XRD, FESEM, DLS and Zeta potential to analyze their physicochemical properties, shape, and stability. The nanoparticles were evaluated for antioxidant, anti-inflammatory, and antimicrobial efficacy, including MIC, MBC, and biofilm inhibition assays in *Vibrio harveyi*, *V. parahaemolyticus*, and *V. alginolyticus*. Toxicity assessment using *Artemia nauplii* confirmed biocompatibility, demonstrating the potential of the nanoparticles for biological applications.

KEYWORDS: *Nanoparticles, Phyllanthus amarus, Acalypha indica, Eclipta prostrata*, *Antibacterial, Anti-biofilm.*